India-Japan Workshop on Biomolecular Electronics and Organic Nanotechnology for Environment Preservation

(IJWBME 2011)

Journal of Physics: Conference Series Volume 358

Himeji, Hyogo, Japan
7-10 December 2011

ISBN: 978-1-62276-775-5
ISSN: 1742-6588

Printed from e-media with permission by:

Curran Associates, Inc.
57 Morehouse Lane
Red Hook, NY 12571

CURRAN ASSOCIATES INC. proceedings.com

Some format issues inherent in the e-media version may also appear in this print version.

Printed by Curran Associates, Inc. (2013)

Institute of Physics
Dirac House, Temple Back
Bristol BS1 6BE UK

Phone: 44 1 17 929 7481
Fax: 44 1 17 920 0979

techtracking@iop.org

Additional copies of this publication are available from:

Curran Associates, Inc.
57 Morehouse Lane
Red Hook, NY 12571 USA
Phone: 845-758-0400
Fax: 845-758-2634
Email: curran@proceedings.com
Web: www.proceedings.com

TABLE OF CONTENTS

Preface

The 'India–Japan Workshop on Biomolecular Electronics & Organic Nanotechnology for Environment Preservation' (IJWBME 2011) will be held on 7–10 December 2011 at EGRET Himeji, Himeji, Hyogo, Japan. This workshop was held for the first time on 17–19 December 2009 at NPL, New Delhi. Keeping in mind the importance of organic nanotechnology and biomolecular electronics for environmental preservation and their anticipated impact on the economics of both the developing and the developed world, IJWBME 2009 was jointly organized by the Department of Biological Functions, Graduate School of Life Sciences and Systems Engineering, the Kyushu Institute of Technology (KIT), Kitakyushu, Japan, and the Department of Science & Technology Centre on Biomolecular Electronics (DSTCBE), National Physical Laboratory (NPL).

Much progress in the field of biomolecular electronics and organic nanotechnology for environmental preservation is expected for the 21st Century. Organic optoelectronic devices, such as organic electroluminescent devices, organic thin-film transistors, organic sensors, biological systems and so on have especially attracted much attention. The main purpose of this workshop is to provide an opportunity for researchers interested in biomolecular electronics and organic nanotechnology for environmental preservation, to come together in an informal and friendly atmosphere and exchange technical knowledge and experience. We are sure that this workshop will be very useful and fruitful for all participants in summarizing the recent progress in biomolecular electronics and organic nanotechnology for environmental preservation and preparing new ground for the next generation.

Many papers have been submitted from India and Japan and more than 30 papers have been accepted for presentation. The main topics of interest are as follows:

- Bioelectronics
- Biomolecular Electronics
- Fabrication Techniques
- Self-assembled Monolayers
- Nano-sensors
- Environmental Monitoring
- Organic Devices
- Organic Functional Materials

We would like to express our sincere thanks to the organizing committee members of this workshop and the many organizations such as the Japan Society for the Promotion of Science (JSPS), Japan, the Department of Science & Technology (DST), India, the Society of Organic Nanometric Interfacial Controlled Electronic (NICE) Devices, the Japan Society of Applied Physics, Himeji City, Himeji Convention & Visitors Bureau, Delhi Technological University, Delhi, India and the University of Hyogo for their financial support. Thanks are also given to The Japan Society of Applied Physics, Division of

Molecular Electronics and Bioelectronics, The Japan Society of Applied Physics (M & BE), the Technical Committee on Dielectric and Electrical Insulation Materials of the Institute of Electrical Engineering in Japan (IEEJ), the Technical Group on Organic Molecular Electronics, Electronics Society of the Institute of Electronics, Information and Communication Engineers (IEICE), and the IEEE Dielectrics and Electrical Insulation Society, Japan Chapter, for their cooperation.

Finally, we hope that the many young and active researchers who are participating will enjoy stimulating discussions and exchange ideas with each other at IJWBME 2011, Himeji, Japan.

IJWBME 2011 Chairs
7 April 2011

Mitsuyoshi Onoda
Graduate School of Engineering, University of Hyogo, Himeji, Japan

Bansi D Malhotra
Department of Biotechnology, Delhi Technological University, Delhi, India

Caption: Participants of the India–Japan Workshop on Biomolecular Electronics & Organic Nanotechnology for Environment Preservation 2011, December 7–10 2011, EGRET Himeji, Japan

Japan Society for the Promotion of Science (JSPS), Japan

Department of Science & Technology (DST), India

Centre on Biomolecular Electronics, and Department of Science & Technology National Physical Laboratory (Council of Scientific & Industrial Research) India

Delhi Technological University, India

The Society of Organic NICE Devices
The Japan Society of Applied Physics

University of Hyogo, Japan

Himeji City, Japan

Himeji Convention & Visitors Bureau, Japan

Technical Committee on Dielectric and Electrical Insulation Materials
The Institute of Electrical Engineers of Japan

Division of Molecular Electronics and bioelectronics
The Japan Society of Applied Physics

Technical Group on Organic Molecular Electronics, The Institute of Electronics Information and Communication Engineers

IEE Dielectrics and Electrical Insulation Society Japan Chapter

Peer review statement

All papers published in this volume of *Journal of Physics: Conference Series* have been peer reviewed through processes administered by the proceedings Editors. Reviews were conducted by expert referees to the professional and scientific standards expected of a proceedings journal published by IOP Publishing.

Aptamer based electrochemical sensor for detection of human lung adenocarcinoma A549 cells

Rachna Sharma[1,3], Ved Varun Agrawal[1], Pradeep Sharma[2], R Varshney[2], R K Sinha[3] and B D Malhotra[4]

[1]Biomedical Instrumentation Section, National Physical Laboratory, Dr. K.S. Krishnan Marg, New Delhi 110012, India

[2]Institute of Nuclear Medicine & Allied Sciences, Lucknow Road, Timarpur, Delhi 110 054, India

[3]Department of Applied Physics, Delhi Technological University, Main Bawana Road, Delhi 110042, India

[4]Department of Biotechnology, Delhi Technological University, Main Bawana Road, Delhi 110042, India

E-mail: bansi.malhotra@gmail.com or bansi.malhotra@dce.ac.in

Abstract. We report results of the studies relating to development of an aptamer-based electrochemical biosensor for detection of human lung adenocarcinoma A549 cells. The aminated 85-mer DNA aptamer probe specific for the A549 cells has been covalently immobilized onto silane self assembled monolayer (SAM) onto ITO surface using glutaraldehyde as the crosslinker. The results of cyclic voltammetry and differential pulse voltammetry studies reveal that the aptamer functionalized bioelectrode can specifically detect lung cancer cells in the concentration range of 10^3 to 10^7 cells/ml with detection limit of 10^3 cells/ml within 60 s. The specificity studies of the bioelectrode have been carried out with control KB cells. No significant change in response is observed for control KB cells as compared to that of the A549 target cells.

1. Introduction

Lung cancer is currently a major cause of cancer related deaths both in the developed and developing nations[1-3]. To reduce high mortality rate arising due to lung cancer, an early diagnosis of the disease is essential. Most of the existing methods presently being used for cancer detection are X-ray, spiral computed tomography[4], sputum cytology[5], positron emission tomography[6], virtual bronchoscopy[7] etc .These methods are based on morphological criteria that cannot be used for early detection of this important disease. Besides this, these are known to be non-specific for cancer classification. Since cancer results from the accumulation of a variety of genetic events (e.g., mutations, rearrangements, and deletions)[8-12] controlling cell growth and differentiation, the studies on these changes may perhaps serve as diagnostically useful molecular markers[13-15]. Development of molecular probes that

[4] To whom any correspondence should be addressed.

can recognize molecular abnormalities during the lung carcinogenesis process is likely to enable clinicians for diagnosis of cancer at early stage and identify its subtypes.

Development of noninvasive tests has been the focus of cancer research that may facilitate earlier diagnosis and treatment of lung cancer. Several studies on variation of the surface specific antigens to differentiate between normal and cancer cells have been performed using antibodies[16]. These surface [1]

specific antigens cannot be exclusively expressed for any single type of cells and may result in false-positivity. To address to this problem, aptamers have emerged as attractive alternatives to antibodies[17,18]. Aptamers, single-stranded oligonucleotides, have been identified for specific cancer cell recognition by applying a cell-SELEX (cell-based systematic evolution of ligands by exponential enrichment) method[19]. Besides this, these biomolecules can fold into unique three dimensional conformations to bind to selected molecules ranging in size from small organic molecules to whole cells with high affinity and selectivity[20].

Aptamers are known to possess several advantages over other recognition molecules like ease of synthesis and modification, long term stability, high specificity etc which making these biomolecules as potential candidates for development of desired analytical devices[21]. Several studies relating to the fabrication of aptamer based biosensors for the detection of proteins[22], molecules[23], cancer cells[24-27] etc have been performed indicating high affinity and specificity of the aptamer probes towards their respective targets. And the electrochemical method of detection is being preferred over other methods of detection since this technique exhibits high signal to noise ratio resulting in high sensitivity, rapid response and reproducibility of the biosensor[28]. Pan et al have fabricated an aptamer based electrochemical sensor for label-free recognition and detection of leukemia cells[26]. Rodriguez et al have reported an aptamer based biosensor for lysozyme detection using electrochemical impedance spectroscopy[29].

In the present manuscript, we report results of studies relating to fabrication of an electrochemical biosensor for detection of lung cancer cells using aptamers. The aptamer functionalized bioelectrode has been prepared by covalent immobilization of the aptamer probe on self assembled monolayer of silane onto the ITO surface using glutaraldehyde as the crosslinker. The response of the bioelectrode towards the target cancer cells has been investigated using cyclic voltammetry and differential pulse voltammetry.

2. Materials and methods

3-(2-aminoethylamino) propyl trimethoxysilane (AEAPTS), glutaraldehyde and amine-terminated aptamer sequence have been purchased from Sigma-Aldrich. All the reagents are of analytical grade and have been used without further purification. De-ionized water (Milli Q 10 TS) with resistivity >18.2 MΩ-cm has been used for preparing all aqueous solutions. Indium-tin-oxide (ITO) coated glass plates have been obtained from Balzers, UK, (Baltracom 247 ITO, 1.1 mm thick) with sheet resistance and transmittance 25 Ωsq^{-1} and 90%, respectively.
Aptamer probe: NH_2 - 5'- ACG CTC GGA TGC CAC TAC AGG GTT GCA TGC CGT GGG GAG GGG GGT GGG TTT TAT AGC GTA CTC AGC TC ATG GAC GTG CTG GTG AC -3'[30]
Target cells: A549 cells
Control cells: KB cells

2.1. Culture medium for cells

A549 cells derived from human lung adenocarcinoma and KB (control) cells derived from head and neck squamous carcinoma are maintained as monolayer cultures at 37 °C in a humidified CO_2 incubator (5% CO_2, 95% air). A549 and KB cells are maintained in high glucose DMEM supplemented with 10% fetal bovine serum (FBS), HEPES (10 mM) and antibiotics (30 $\mu g ml^{-1}$ penicillin G, 50 $\mu g ml^{-1}$ streptomycin and 2 $\mu g ml^{-1}$ nystatin). All the cell lines are routinely sub-cultured (twice a week) using 0.05 % trypsin in 0.02 % EDTA and reseeded in fresh medium. FBS and DMEM (both low and high glucose) were purchased from SIGMA chemicals.

2.2. Fabrication of Aptamer-functionalized bioelectrode
ITO coated glass plates are washed with acetone, ethanol and copious amount of deionized water followed by hydrolyzation in NH_4OH:H_2O_2:H_2O (1:1:5) solution at 80°C for 45 min. Hydrolyzed ITO plates are dipped in 1% ethanolic solution of AEAPTS for 2 h for formation of self-assembled monolayer (SAM) and then washed with ethanol followed by deionized water. Silanized ITO surfaces are then soaked in 0.5% aqueous solution of glutaraldehyde for 6 h for SAM formation. Glutaraldehyde acts as a crosslinker between the amine group at the silanized ITO surface and the amine-terminated aptamer sequence. 20 µL of 1 µM aptamer solution is then dispensed onto the silanized ITO surface for about 2 h for covalent immobilization resulting in the fabrication of aptamer functionalized bioelectrode.

2.3. Characterization
The morphology of the aptamer functionalized bioelectrode and morphological changes on incubation of the bioelectrode with target cells have been studied using SEM, LEO 440 scanning electron microscope. The transmission studies of the silanized ITO surface and the aptamer functionalized bioelectrode in the infrared region have been carried out on Perkin Elmer, Spectrum BX II spectrophotometer in the wavenumber range of 4000-450 cm^{-1}.

The electrochemical experiments have been conducted on Autolab PGSTAT 302N System (Ecochemie, The Netherlands) in a three electrode system. All electrochemical experiments have been carried out in a cell containing 15 ml of 100 mM Phosphate Buffer Solution (PBS, pH 7.4) containing 0.9% NaCl and 5 mM $K_3/K_4[Fe(CN)_6]$ as a redox probe and using a platinum wire as auxiliary, a Ag/AgCl wire as reference, and the modified ITO as the working electrode.

3. Results and discussion

3.1. SEM studies
Figure 1 shows the morphology obtained for the apt-glu-silane-ITO bioelectrode before and after cell binding. The observed granular structure on immobilization of single-stranded DNA aptamer indicates covalent binding of aminated aptamer probe onto silanized ITO surface (figure 1(a)). The uniform spindle like structure appears on incubation of the apt-glu-silane-ITO bioelectrode with A549 cells (figure 1(b)) revealing the affinity of aptamers with target cells.

Figure 1. SEM micrographs of apt-glu-silane-ITO bioelectrode before (a) and after (b) cell binding.

3.2. FTIR studies
Figure 2 shows the transmittance spectra observed for the silanized ITO electrode and aptamer functionalized bioelectrode. The peaks seen at 679 cm^{-1} and 1113 cm^{-1} corresponding to Si-CH_3 stretch and Si-O-Si antisymmetric stretch respectively, indicate the formation of siloxane bond between the

silane and the ITO surface[31] [figure 2(a)]. Peaks found at 1415 cm^{-1}, 2808 cm^{-1} and 3265 cm^{-1} correspond to C-N, C-H and N-H stretch in the AEAPTS, respectively. Further, the peak at 1585 cm^{-1} corresponding to NH_2 deformation reveals the presence of amine groups at the silanized ITO surface.

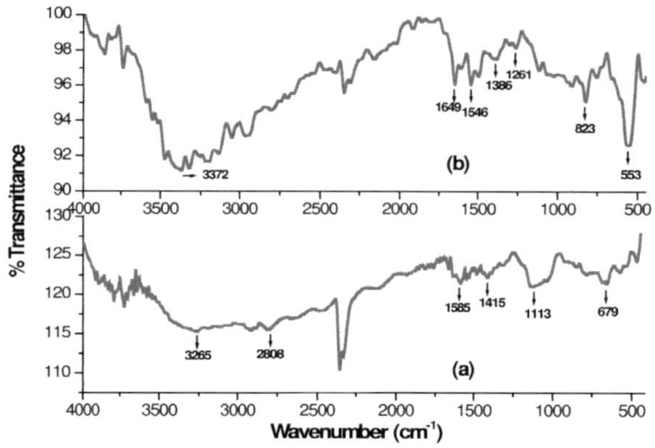

Figure 2. FTIR spectra of (a) AEAPTS/ITO surface and; (b) Apt/Glu/AEAPTS/ITO surface.

In figure 2(b), peaks at 553 cm^{-1} and 823 cm^{-1} correspond to C-O and N-H bending vibrations in amino acids, respectively. Peaks seen at 3372 cm^{-1} and 1386 cm^{-1} correspond to N-H stretch and C-N stretch, respectively. Further, the peak at 1261 cm^{-1} corresponds to P-O stretch of the phosphate backbone of DNA, peaks at 1546 cm^{-1} and 1649 cm^{-1} pertain to the C-O stretch of the purine and pyrimidine rings of DNA[32]. All these peaks reveal the immobilization of the amine-terminated aptamer probe onto the silanized ITO electrode.

3.3. Electrochemical characterization

Figure 3a shows the cyclic voltammograms obtained for the ITO electrode, silane-ITO electrode, glu-silane-ITO electrode and apt-glu-silane-ITO bioelectrode in the potential range of -0.7V to +0.7V at scan rate of 50 mV/s. The 0.4V peak corresponds to the oxidation peak of the redox couple, K_3/K_4 $[Fe(CN)_6]$ present in the buffer. It can be seen that there is a significant increase in the anodic peak current for the silane-ITO electrode (figure 3a (ii)) compared to that of the ITO electrode (figure 3a (i)). The NH_2-terminal of the silane acquires positive charge in the buffer and thus attracts the negatively charged cyanide ions present in the buffer. The electrostatic attraction between the positively charged silanized ITO surface and the negatively charged redox species present in the buffer facilitates electron transfer across the electrode-electrolyte interface leading to increased oxidation current[33]. Futher, decrease in oxidation current may perhaps be attributed to the formation of an insulating layer of glutaraldehyde on the silanized ITO surface (figure 3a (iii)). Again, the increase in the oxidation current on immobilization of the amine-terminated aptamer probe (figure 3a (iv)) can perhaps be asisgned to the electrostatic attraction between the positively charged surface of apt-glu-silane-ITO bioelectrode and the negatively charged redox species present in the buffer resulting in facile transfer of electrons.

Figure 3b shows the differential pulse voltammograms obtained for the ITO electrode, silane-ITO electrode, glu-silane-ITO electrode and apt-glu-silane-ITO bioelectrode in the potential range of -0.7V to +0.7V. The increase in the peak current for silane-ITO electrode (figure 3b (ii)) compared to that of the ITO electrode (figure 3b (i)) is attributed to the facile electron transfer aided by the opposite charges on electrode and the electrolyte. The decrease in the peak current for glu-silane-ITO electrode (figure 3b (iii)) reveals the formation of an insulating layer of glutaraldehyde. Further, increase in the current on immobilization of the amine-terminated aptamer sequence (figure 3b (iv)) is ascribed to the transfer of electrons facilitated due to the electrostatic attraction between positive electrode surface and negative redox species in the buffer.

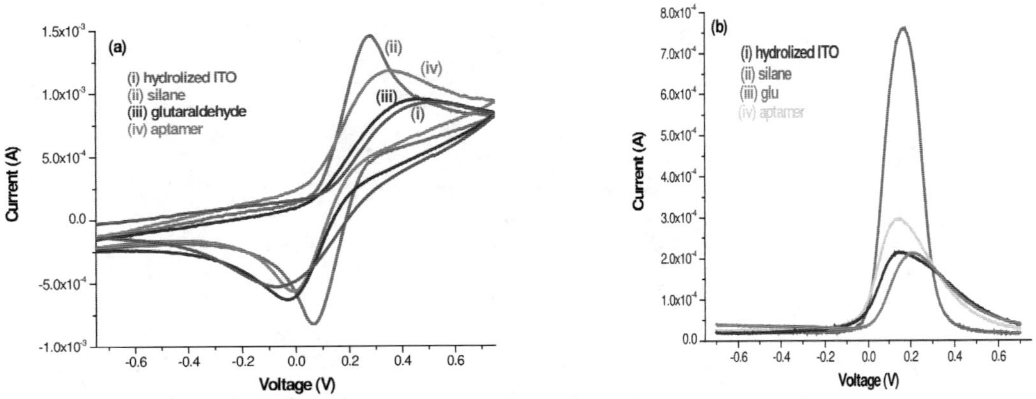

Figure 3. Stepwise characterization of aptamer functionalized bioelectrode using CV (a) and DPV (b).

3.4. Surface coverage of aptamer

The cyclic voltammteric investigations of apt-glu-silane-ITO bioelectrode have been conducted as a function of scan rate (v) (10–100 mV/s) to determine the surface coverage of aptamer probe. The anodic peak potential (E_{pa}) is found to vary linearly with logarithm of scan rate and follows Eq.1:

$E_{pa} = 0.2051 \log v - 0.0433$; R = 0.9949; SD = 0.0046 Eq.1

According to Laviron's theory, the slope of the linear curve between the anodic peak potential and the logarithm of scan rate represents RT/αnF (α- transfer coefficient). This can be used to calculate the surface concentration of the aptamer using the following equation:

$i_p = n^2F^2vCA\,(4RT)^{-1}$ Eq.2

where, i_p/v can be calculated from the i_p Vs. v plot (i_p- anodic peak current; v- scan rate)[34].

The total surface concentration of the aptamer probe has been found to be 1.627×10^{-10} mol cm^{-2} (using $i_p/v = 9.754 \times 10^{-6}$).

3.5. Response studies of aptamer functionalized bioelectrode

The response of the aptamer functionalized bioelectrode towards target cell (A549) binding has been investigated using cyclic voltammetry in PBS buffer containing $K_3/K_4[Fe(CN)_6]$ as a redox couple in the voltage range of -0.7 V to +0.7 V at scan rate of 50 mV/s. The decrease in the oxidation current is observed due to binding of the A549 cells (figure 4a). This is because the biological cells on attachment to the substrate form an insulating layer that perhaps provides barrier to the flow of electrons through the electrode surface resulting in the decrease of the current[26]. The resistance to the flow of electrons increases with increasing concentration of A549 cells and thereby results in the decrease of current with rising concentration of target cells.

The calibration curve for the apt-glu-silane-ITO bioelectrode can be obtained by plotting the anodic peak current for logarithmic values of the different concentrations of cholesterol (figure 5(a)). Linear relation is obtained between the anodic peak current and logarithmic value of cell concentration over the range of 10^3-10^7 cells/ml with standard deviation and correlation coefficient of 62.91 μA and 0.9831, respectively.

The binding of target cells to the apt-glu-silane-ITO bioelectrode has been studied using differential pulse voltammetry. The decrease in current is observed with the increasing concentrations of target cells (figure 4b) due to formation of the insulating layer which impedes the flow of electrons through the electrode surface indicating binding of the cells to the aptamer. The calibration curve obtained by plotting the DPV peak current for the logarithmic values of cholesterol concentrations (figure 5(b)) reveals the linearity over the concentration range of 10^3-10^7 cells/ml with standard deviation and correlation coefficient of 13.38 μA and 0.9863, respectively.

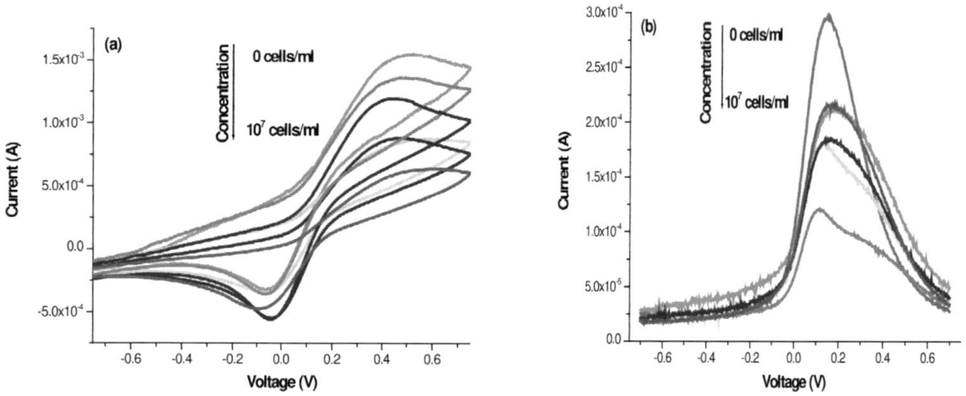

Figure 4. Response Studies of the aptamer functionalized bioelectrode using CV (a) and DPV (b) with target A549 cell concentrations of (i) 0; (ii) 10^3; (iii) 10^4 ; (iv) 10^5 ; (v) 10^6 and; (vi) 10^7 cells/ml.

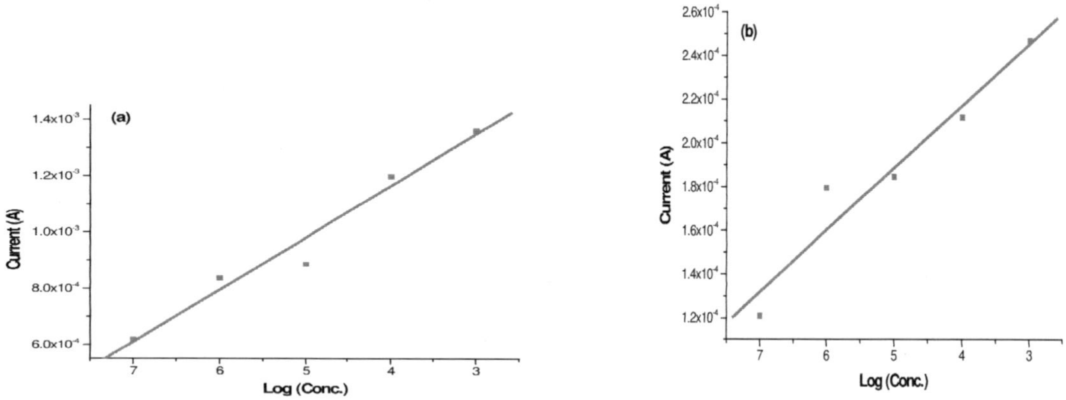

Figure 5. Calibration plot for CV response (a) and DPV response (b) of the aptamer functionalized bioelectrode towards different concentrations of A549 cells.

3.6. Response studies of aptamer functionalized bioelectrode

The specificity of the apt-glu-silane-ITO bioelectrode towards detection of A549 cells has been investigated using cyclic voltammetry (figure 6a). In the control experiments, the response of the bioelectrode towards culture medium and the control KB cells has been studied. The observed decrease in the current observed on incubation of the bioelectrode with culture medium without cells may be attributed to the physical adsorption of the species from the environment. The decrease in the current on incubation of the bioelectrode with KB cells is approximately half of that in case of A549 cells for the same concentration. Thus, the binding of the A549 cells to the bioelectrode surface perhaps dominates the CV response.

The selective response of the apt-glu-silane-ITO bioelectrode towards the A549 cells has been further confirmed by differential pulse voltammetry (figure 6b). The decrease in signal observed with the target A549 cells is much higher as compared to that obtained with culture medium and the control

KB cells. This further reveals the selectivity of the aptamer functionalized bioelectrode towards human lung adenocarcinoma A549 cells.

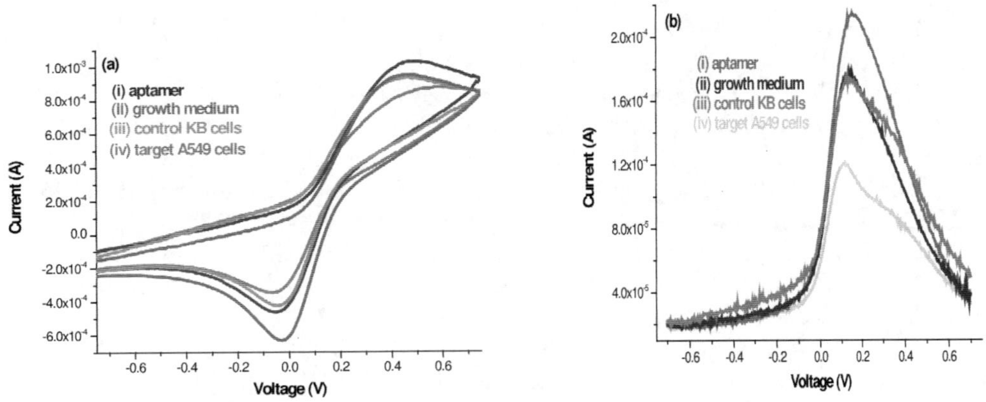

Figure 6. Specificity studies of the aptamer functionalized bioelectrode with control KB cells using CV (a) and DPV (b).

4. Conclusions

An aptamer based biosensor has been fabricated for detection of human lung adenocarcinoma A549 cells using aminosilane as the platform. The amine-terminated aptamer probe has been covalently bonded to the amine functionalized silane surface onto ITO using glutaraldehyde as the crosslinker. The surface coverage of the aptamer probe onto the silanized ITO electrode is found to be 1.627×10^{-10} mol cm^{-2}. This electrochemical aptamer sensor exhibits detection limit of 10^3 cells/ml with response time of 60 s for the target A549 cells. Aptamers show selectivity towards their target cells as the electrochemical signal response for control KB cells is about 50% as compared to that of the target cells. Thus aptamers can be utilized for the detection of the lung cancer cells. It should be interesting to utilize conducting polymers and nanomaterials including nanostructured metal oxides etc [35,36] to obtain increased sensitivity of this aptasensor.

References

[1] Kumar M, Agarwal S K and Goel S K 2009 *Mol. Cell Biochem.* **322** 73
[2] Pathak A K, Bhutani M, Mohan A, Guleria R, Bal S and Kochupillai V 2004 *Indian J. Chest Dis. Allied Sci.* **46** 191
[3] Behera D and Balamugesh T 2004 *Indian J. Chest Dis. Allied Sci.* **46** 269
[4] Sutedja G 2003 *Eur. Respir. J.* **21** 57s
[5] Mao L, Hruban R H, Boyle J O, Tockman M and Sidransky D 1994 *Cancer Res* **54** 1634
[6] Weder W, MD, Schmid R A, Bruchhaus H, Hillinger S, Schulthess G K and Steinert H C 1998 *Ann. Thorac Surg.* **66** 886
[7] Schmiemann V, Bocking A, Kazimirek M, Onofre A S C,Gabbert H E, Kappes R, Gerharz C D and Grote H J 2005 *Clin Cancer Res* **11** 7728
[8] Zhu H, Smith C, Ansah C and Gooderham N J 2005 *Cancer Cell International* **5** 28
[9] Mitsudomi T, Steinberg S M, Oie H K, Mulshine J L, Phelps R, Viallet J, Pass H, Minna J D and Gazdar A F 1991 *Cancer Res.* **51** 4999
[10] Wu W, Fan Y H, Kemp B L, Walsh G and Mao L 1998 *Cancer Res.* **58** 4082
[11] Lynch T J,Bell D W,Sordella R, Gurubhagavatula S, Okimoto R A, Haluska F G, Louis D N, Christiani D C, Settleman J and Haber D A 2004 *N. Engl. J. Med.* **350** 21
[12] Croce C M 2008 *N. Engl. J. Med.* **358** 502
[13] Li Y, Lee H J and Corn R M 2007 *Anal. Chem.* **79** 1082

[14] Li Z, Wang Y, Wang J, Tang Z, Pounds J G and Lin Y 2010 *Anal. Chem.* **82** 7008
[15] Wang H, Cao C, Li B, Chen S, Yin J, Shi J, Ye D, Tao Q, Hu P, Epstein A and Ju D 2008 *Cancer Immunol Immunother* 57 677
[16] Lee T Y and Song J T 2010 *Thin Solid Films* **518** 6630
[17] Syed M A and Pervaiz S 2010 *Oligonucleotides* **20** 215
[18] Cho E J, Lee J W and Ellington A D 2009 *Annu. Rev. Anal. Chem.* **2** 241
[19] Guo K T, Paul A, Schichor C, Ziemer G and Wendel H P 2008 *Int. J. Mol. Sci.* **9** 668
[20] Mok Wand Li Y 2008 *Sensors* **8** 7050
[21] Proske D, Blank M, Buhmann R and Resch A 2005 *Appl. Microbiol. Biotechnol.* **69** 367
[22] Ishii Y, Tajima S and Kawarada H 2011 *App. Phys. Exp.* **4** 027001
[23] Prabhakar N, Matharu Z, Malhotra B D 2011 *Biosensors and Bioelectronics* **26** 4006
[24] Liu G, Mao X, Phillips J A, Xu H, Tan W and Zeng L 2009 *Anal. Chem.* **81** 10013
[25] Jie G, Wang L, Yuan J and Zhang S 2011 *Anal. Chem.* **83** 3873
[26] Pan C, Guo M, Nie Z, Xiao X, Yao S 2009 *Electroanalysis* **21** 1321
[27] Pan Y, Guo M, Nie Z, Huang Y, Pan C, Zeng K, Zhang Y and Yao S 2010 *Biosensors and Bioelectronics* **25** 1609
[28] Willner I and Zayats M 2007 *Angew. Chem. Int. Ed.* **46** 6408
[29] Rodriguez M C, Kawde A N and Wang J 2005 *Chem. Commun.* 4267
[30] Zhao Z, Xu L, Shi X, Tan W, Fang X and Shangguan D 2009 *Analyst* **134** 1808
[31] Das M, Sumana G, Pandey M K, Nagarajan R and Malhotra B D 2011 *Sensor Letters* **9** 499
[32] Das M, Dhand C, Sumana G, Srivastava A K, Nagarajan R, Nain L, Iwamoto M, Manaka T and Malhotra B D 2011 *Biomacromolecules* **12** 540
[33] Arya S K, Prusty A K, Singh S P, Solanki P R, Pandey M K, Datta M, Malhotra B D 2007 *Anal. Biochem.* doi:10.1016/ j.ab.2007.01.029
[34] Singh R, Dhand C, Sumana G, Verma R, Sood S, Gupta R K and Malhotra B D 2010 *J. Mol. Recognit.* DOI:10.1002/jmr.1014
[35] Huang Y F, Lin Y W and Lin Z H 2009 *Nanopart. Res.* **11** 775
[36] Lin M M, Kim D K, Haj A J and Dobson J 2008 *IEEE Trans. Nanobiosci.* **7** 298

Stability of polypyrrole soft actuators in ionic liquids

Keiicgi Kaneto[1] **and Wataru Takashima**[2]
[1]Graduate School of Life Science and Systems Engineering, Kyushu Institute of Technology, 2-4 Hibikino, Wakamatsu-ku, Kitakyushu 808-0196, Japan
[2]Research Center for Advanced Eco-fitting Technology, Kyushu Institute of Technology, 2-4 Hibikino, Wakamatsu-ku, Kitakyushu 808-0196, Japan

E-mail: kaneto@life.kyutech.ac.jp

Abstract. Characteristics of electrochemomechanical deformation (ECMD) of polypyrrole films using ionic liquids are reported. The PPy film prepared by electrodeposition in an ionic liquid (1-Butyl-1-methylpyrrolidinium bis(trifluorometylsulfonyl)imide, BMPTFSI) was compact and high density. The other film prepared from LiTSFI/methyl benzoate and dimethyl phthalate mixed solvents was porous and low density. Both films demonstrated a stable ECMD in the ionic liquid. The strain of ECMD was 3-5% and superimposed on a creeping, showing a typical behaviour of cation movement. The Strains of ECMD in both films operated in a mixed electrolyte of BMPTFSI and propylene carbonate were enhanced up to 17- 25 %, showing anion movement. However, the large strain decreased upon several electrochemical cycles. The results were discussed in terms of swelling of the PPy film by solvents and loss of electrochemical activity.

1. Introduction

Electrochemomechanical deformation (ECMD) of conducting polymers[1] are interested in application to artificial muscles or soft actuators for robots, since the strain and the stress have been achieved more than 30% and 22MPa[2,3], respectively. These figures are comparable or even better than those of typical skeletal muscles. The ECMD is mainly generated by insertion of counter ions into polymer matrix as well as by the conformation change of polymer backbone due to the delocalization of π-electron upon oxidation. However, the short cycle life, which results mostly from the degradation of π-electron system[4], has been the issue to be solved for the practical use.

Recently, ionic liquids (ILs), which show non flammability, non volatility, high ionic conductivity and wide potential window, are interested and employed as electrolytes for electrochemical reactions [5-8]. Few studies on ECMD of conducting polymers prepared from ionic liquids are found in literatures. Characteristics of ECMD or actuation of conducting polymers[7-14], polypyrrole (PPy)[8] polyaniline[9] and poly(3-methythiophene)[10] in ionic liquids have been reported. These materials were prepared electrochemically using conventional electrolyte solutions[11-13,14] and chemical polymerization[9]. The ionic liquids employed commonly were 1-ethyl-3-methylimidazolium (EMI^+) and 1-buthyl-3-methylimidazolium (BMI^+) as cations. As anions trifluoromethansulfonic ($CF_3SO_3^-$), bis(trifluorometylsulfonyl)imide ($TFSI^-$) and PF_6^- were used. The chemical structures of ionic liquids and materials used in this study are depicted in Fig.1.

[1] To whom any correspondence should be addressed.

The ECMD of these polymers in ionic liquids demonstrated excellent cycle stability even more than million times[9,11-13], compared with those cycled in conventional electrolyte[2-4]. However, the overall performances of ECMD in ionic liquids were not satisfactorily regarding with the strain, which was less than ca. 3% along the film length[12]. Possibly, this is a reason why the ECMD using ionic liquid has not been studied intensively, so far. Recently, 11% strain of thickness direction (estimated by ellipsometry) in PPy films cycled in EMITSFI was reported[14]. The result indicated the anisotropic deformation of the PPy film during the electrochemical cycles.

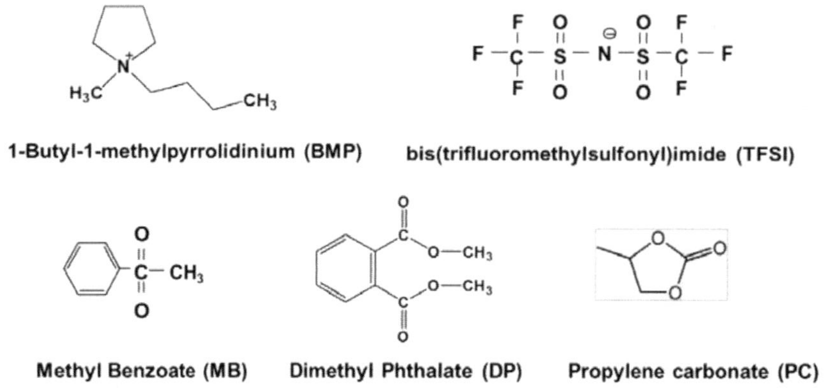

Figure 1. Chemical structures of ILs and used materials.

PPy films prepared from LiTFSI/methyl benzoate (MB) was low density and high porosity in the surface morphology[3], demonstrating the world record of largest strain, viz. 39%. The study motivated us to see the enhanced strain and increased stability of ECMD by using porous PPy films and ILs. In this paper, characteristics of ECMD in PPy films electrodeposited from ILs of BMPTFSI and LiTFSI/MB operated in the IL is reported, comparing with the results operated in the mixture of IL and propylene carbonate (PC).

2. Experimental

Pyrrole, LiTFSI, MB, dimethyl phthalate (DP) and PC were purchased from Wako Pure Chemical Industries Ltd., and used without further purification. IL of BMPTFSI was obtained from Fuluka Co., Ltd. PPy films were prepared by two methods of electro polymerizations of pyrrole from #1: pure BMPTFSI and #2: mixed electrolyte solution of 0.2 M LiTFSI/MB&DP (=1/5). The concentrations of pyrrole in solutions were 0.1M for #1 and 0.25 M for #2, and these polymerizations were carried out at constant current of 0.2 mA/cm^2 for 6 h at 0℃.

The test piece of films was 15 x 2 mm and set in a electrochemical cell for the measurement of ECMD same to that in our previous papers[2-4]. The electrolytes for ECMD measurement were the pure BMPTFSI and the mixture of BMPTFSI and PC by the ratio of 1:1. A Ag/AgCl wire was used as a pseudo-reference electrode[2-4,8]. Measurements of cyclic voltammetry (CV) and strain of ECMD were carried out at the ambient condition. The strain was defined by $\Delta l/l_0$, where $\Delta l = l - l_0$, l and l_0 were the film lengths at the time measured and the original, respectively.

3. Results and discussion

Table 1 shows characteristics of the obtained films containing TFSI as the dopant for their densities, conductivities (σ) and film morphologies by scanning electron micrograph (SEM). As seen in SEM the morphologies of films were apparently different between #1 and #2. Namely, the #1 film was dense and compact compared with #2 film being porous and rough surface, which were consistent with their densities and conductivities. The thickness of films was ca. 20 and 40 μm for #1 and #2 films, respectively.

Figure 2(a) depicts the CV and strain of ECMD in compact #1 film in the IL of BMPTFSI at the potential between -1.5V and +1.3V and the scan rate of 2mV/s. The film showed the cyclic strain, $\Delta l/l_0$ of 3-5 %, responding to oxidation and reduction. The cyclic strain was superimposed on a creep up to 10 %. The expansion of the film by reduction indicates the cation movement and insertion into the film[15]. The similar results of cation movement have been also reported by other groups[9-12,14], indicating immobilized TFSI in the film. The current peak of oxidation shifted to higher potentials without dropping the magnitude, and the reduction peak was almost unchanged. On the other hand, in the mixed solution of BMPTFSI/PC as shown in Fig.2 (b), the cyclic strain increased by oxidation, indicating the anion movement and insertion. It is noted that the cyclic strain was more than 17% at the beginning, however, faded out into the creep of 7% in a few cycles. The oxidation peak in the CV shifted to higher potentials and decreased in the magnitude. The reduction current also decreased similarly. The results indicate that the film apparently degraded in the activity upon electrochemical cycling.

Table 1. Scanning electron micrographs of PPyTFSI films obtained from #1IL and #2 mixed solution of MB&DP

PPyTFSI Film			Surface (Scale 10 μm)		Cross section (Scale: 20 mm)
	Density (g/cm³)	σ (S/cm)	Liquid side	Electrode side	
#1 IL	1.67	99			
#2 MB& DP	0.89	57	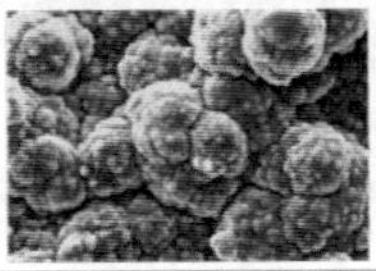	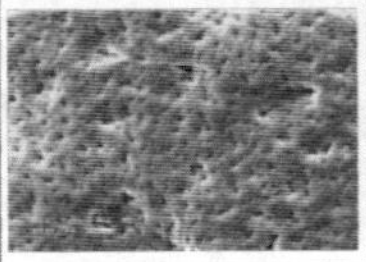	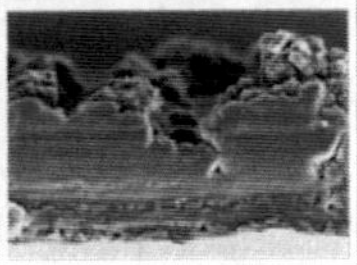

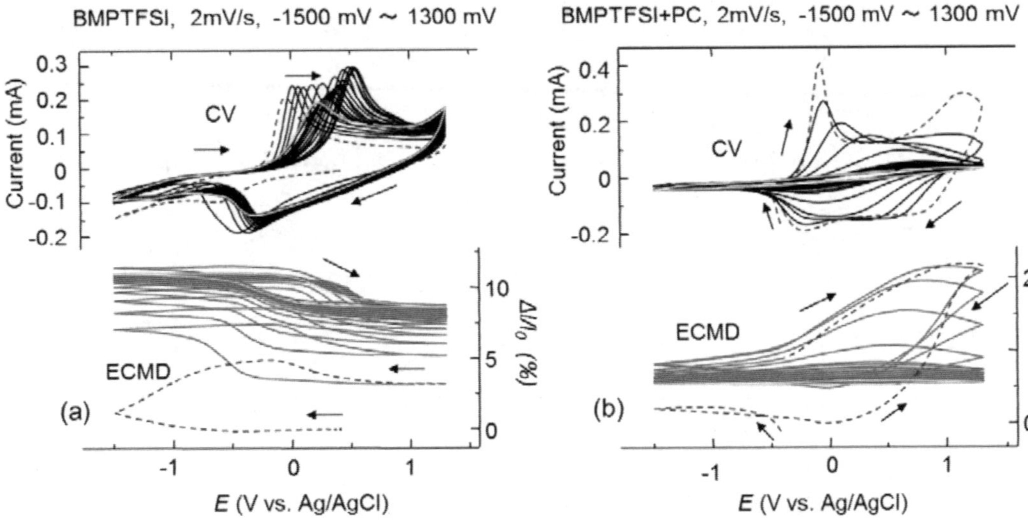

Figure 2. CVs and ECMD strains in #1 PPyTFSI/IL film operated in (a) Ionic liquid of BMPTFSI and (b) mixed BMPTFSI/PC (=1/1) electrolytes. Broken lines are the first cycle.

Figure 3 (a) shows the CV and ECMD of porous #2 film operated in the pure ionic liquid, indicating similar behaviour to #1 film except for the magnitude of cyclic strain and the small sift of oxidation current peak. The cyclic strain was about 5% and was kept almost constant upon many cycles. On the other hand as depicted in Fig.3 (b) in the mixed electrolyte of BMPTFSI/PC, it is noted that the magnitude of cyclic strain was enhanced to be more than 25 %. However, by cycling the strain gradually faded out to the creep of 16% similarly to that of Fig.2 (b).

Figure 4 (a) and (b) summarize cyclic stabilities of ECMD, Δl of #1 and #2 films as the function of charge for the pure IL and the mixed electrolyte for comparison. The charge is estimated from the total consumed electric charges during cycling, and the negative direction corresponds to reduction. The shift to negative direction in the charge indicates over reduction due to the unbalanced electrochemical cycles. In other words, the smaller shift implies the better cycle stability. Upward shift until levelling off is due to a creep, in this case, swelling of films by sorption of ionic liquid and solvents as discussed in our previous paper[15]. The shift was larger in porous #2 film than that of compact #1 film.

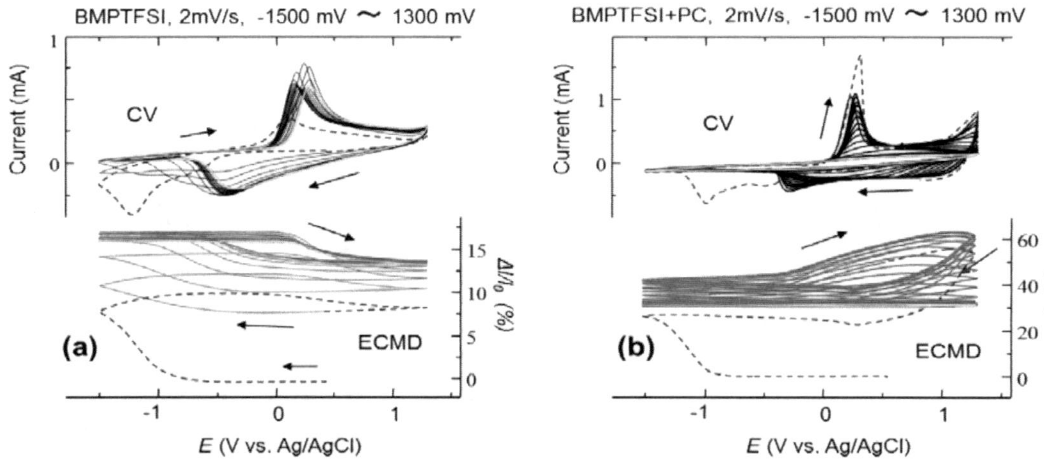

Figure 3. CVs and ECMD strains in #2 PPyTFSI/MB&DP film operated in (a) Ionic liquid of BMPTFSI and (b) mixed BMPTFSI/PC (=1/1) electrolytes. Broken lines are the first cycle.

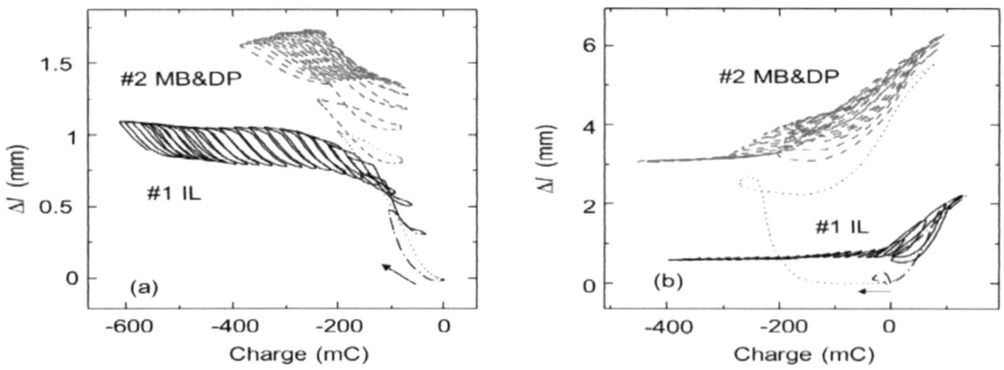

Figure 4. Deformation, D*l* of PPy/TFSI film operated in (a) Ionic liquid of BMPTFSI and (b) BMPTFSI/PC mixed solution for #1 (lower curves) and #2 (upper curves) films.

Figure5 shows the stability of cyclic strain as the function of electrochemical cycle. It is apparent that the cyclic strains in #1 and #2 films behaved similar manner in the same electrolyte solutions. Again the porous film exhibited larger strain compared with that obtained from the IL by 1.5-2 times at the beginning of cycling, then, the strains resulted in the similar magnitude. The cyclic strains of both films in the mixed solvent of IL/PC were larger than those in IL. It is conjectured that by use of the mixed solution the PC swollen the films and accommodated with TFSI, resulting in TFSI anion insertion and the large strain. The shrinking of cyclic strain was resulted from the degradation of films, being due to lose of the conductivity[4]. This was confirmed by the fact that the conductivity of #2 film after the ECMS measurement was 1.7 S/cm. The degradation of films in the mixed solution may be initiated by the over oxidation of π-conjugation with the existence of PC, as supposed from the variation of CV curves in Fig.2 and Fig.3. The detail mechanisms for the degradation in the mixed solution may be complicated and not be discussed at the present stage. We are expecting, however, that the performance of ECMD in pure ILs could be improved by controlling the film morphology and tuning the operation potentials appropriately, as well as choosing suitable ILs.

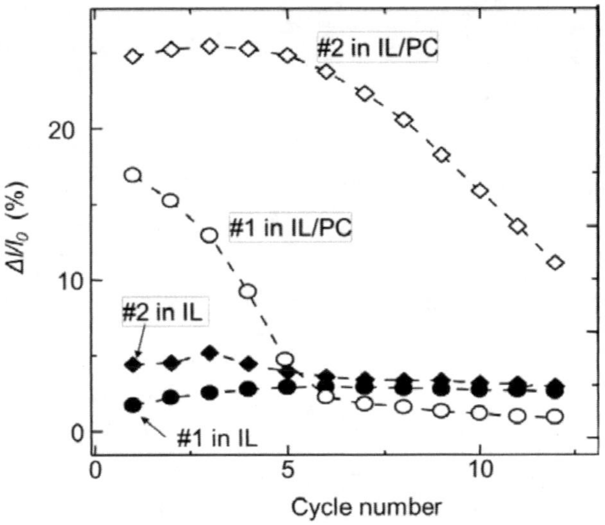

Figure5. ECMD of films #1 and #2 in IL, BMPTFSI (solid symbols) and BMPTFSI/PC mixed solution (open symbols).

Acknowledgments

This work was supported by a Grant-in-Aid for Science Research from the Ministry of Education, Culture, Sports, Science and Technology, Japan and the centre for advanced ecological techniques of Kyushu Institute of Technology. The authors acknowledge to Dr. Kentaro Yamato for his efforts to carry out the experiment of the work.

References

[1] Mirfakhrai T, Madden J DW and Baughman R H 2007 *Materials Today* **10** 30
[2] Hara S, Zama T, Takashima W and Kaneto K 2004 *Polym. J.* **36** 151
[3] Hara S, Zama T, Takashima W and Kaneto K 2005 *Smart Mater. Struct.* **14** 1501
[4] Kaneto K, Suematsu H and Yamato K 2008 *Advances in Science and Technology,* **61** 12
[5] Sekiguchi K, Atobe M and Fuchigami T 2002 *Electrochem. Commun.* **4** 881
[6] Sekiguchi K, Atobe M and Fuchigami T 2003 *J. Electroanal. Chemistry* **557** 1
[7] Lu W, Fadeev A G, Qi B and Mattes BR 2003 *Synth. Metals* **135-136** 139
[8] Pringle J M, Efthimiadis J, Howlett P C, Efthimiadis J, MacFarlane D R, Chaplin A B, Hall S B, Officer DL, Wallace G G and Forsyth M 2004 *Polymer* **45** 1447

[9] Lu W, Fadeev A G, Qi B, Smela E, Mattes B R, Ding J, Spinks G, Mazukiewizc J, Zhou D, Wallace G G, MacFarlane D R, Forsyth S A, Forsyth M 2002 *Science*, **297** 983
[10] Xi B, Truong V -T, Whitten P, Ding J, Spinks G M and Wallace G G 2006 *Polymer* **47** 7720
[11] Ding J, Zhou D, Spinks G M, Wallace G G, Forsyth S, Forsyth M and MacFarlane D 2003 *Chem.Mater* **15** 2392
[12] Spinks G M, Xi B, Zhou D, Truong V –T and Wallace G G 2004 *Synthetic Metals* **140** 273
[13] Randriamahazaka H, Plesse C, Teyssie D and Chevrot C 2005 *Electrochimica Acta*, **50** 4222
[14] Samani M B, Cook C D, Madden J D, Spinks G M and Whitten P G 2006 *Thin Solid Films* **516** 2800
[15] Kaneto K, Shinonome T, Tominaga K and Takashima W 2011 *Jpn. J. Appl. Phys.* **50** 091601

Strategies for an enzyme immobilization on electrodes: Structural and electrochemical characterizations

V Ganesh[1] and A Muthurasu
Electrodics and Electrocatalysis (EEC) Division, CSIR – Central Electrochemical Research Institute (CECRI), Karaikudi – 630006, Tamilnadu, India.

E-mail: ganelectro@gmail.com or vganesh@cecri.res.in

Abstract. In this paper, we propose various strategies for an enzyme immobilization on electrodes (both metal and semiconductor electrodes). In general, the proposed methodology involves two critical steps viz., (1) chemical modification of substrates using functional monolayers [Langmuir – Blodgett (LB) films and/or self-assembled monolayers (SAMs)] and (2) anchoring of a target enzyme using specific chemical and physical interactions by attacking the terminal functionality of the modified films. Basically there are three ways to immobilize an enzyme on chemically modified electrodes. First method consists of an electrostatic interaction between the enzyme and terminal functional groups present within the chemically modified films. Second and third methods involve the introduction of nanomaterials followed by an enzyme immobilization using both the physical and chemical adsorption processes. As a proof of principle, in this work we demonstrate the sensing and catalytic activity of horseradish peroxidase (HRP) anchored onto SAM modified indium tin oxide (ITO) electrodes towards hydrogen peroxide (H_2O_2). Structural characterization of such modified electrodes is performed using X-ray photoelectron spectroscopy (XPS), atomic force microscopy (AFM) and contact angle measurements. The binding events and the enzymatic reactions are monitored using electrochemical techniques mainly cyclic voltammetry (CV).

1. Introduction

Self-assembled monolayer (SAM) forms the basis to understand the structural arrangement, electron transfer across the interface and the flexibility of macro-molecules to the surrounding environment, which are the key factors for the realization of biomimics [1]. As a first step towards the development of synthetic biomimetic systems, it is very much important to have a thorough knowledge on enzymes and their activity in terms of their structural arrangement and enzymatic reactions. SAM provides a very good platform for immobilizing enzymes and investigating their redox properties that help in understanding their sensing ability and catalytic reactions. Keeping this in mind, in this work we propose a simple methodology to anchor horseradish peroxidase (HRP) on 3-mercaptopropyltrimethoxysilane (MPTMS) monolayer coated ITO electrodes. Gold nanoparticles (GNPs) are introduced into this sensing system in order to enhance the loading of enzymes and to tune their sensitivity and detection limit. Several reviews are available on enzyme immobilization and structural evaluation [2,3] using a variety of spectroscopic, microscopic and optical techniques.

Determination of hydrogen peroxide (H_2O_2) is of considerable interest in recent times. It is not only a by-product of several highly selective oxidases enzymes but also an essential mediator in food, pharmaceutical, clinical, industrial and environmental analysis [4]. H_2O_2 can be detected at a low applied

[1] To whom any correspondence should be addressed.

Published under licence by IOP Publishing Ltd

potential by using peroxidase enzyme as an electro-catalyst for monitoring the electrochemical reduction process. The peroxidase enzyme contains heme as a prosthetic group, which is also the protein active site with the resting state of heme iron as Fe(III), and it can catalyze the H_2O_2 reduction. Some of the substrates were reported for this process [5-9]. However, when HRP is immobilized on an electrode, direct electron transfer between enzyme and the electrode becomes difficult because the electro active sites present within the enzyme are deeply buried by the peptide chains. The immobilized enzyme can also partially lose its bioactivity due to an unfavorable orientation at the electrode surface [10]. Therefore, suitable electrode materials and effective enzyme immobilization methods that are capable of facilitating direct electron transfer while retaining the bioactivity of enzyme at the same time are most crucial factors for the success of a biosensor. Several methods including adsorption [11], cross-linking [12], entrapment in sol-gel [13], layer-by-layer assembly [14] have been used to immobilize HRP enzyme on electrodes. Even then the application and development of biosensors have been limited due to the lack of a simple and stable enzyme immobilization approach to monitor the enzymatic reactions. On the other hand, ITO offers an attractive substrate due to their prominent characteristics such as excellent optical transparency, high electrical conductivity, wide electrochemical working window, excellent substrate adhesion and stable electrochemicalproperties [15-18], which could be exploited for immobilizing enzymes and monitoring their catalytic processes.

In this work, we propose a simple methodology to immobilize HRP on SAM modified ITO electrode to monitor its sensing and catalytic behavior towards H_2O_2 as shown in scheme 1. MPTMS is used for monolayer formation and GNPs are used as a matrix to bind the enzyme effectively and to facilitate direct electron transfer between the enzyme and the electrode surface.

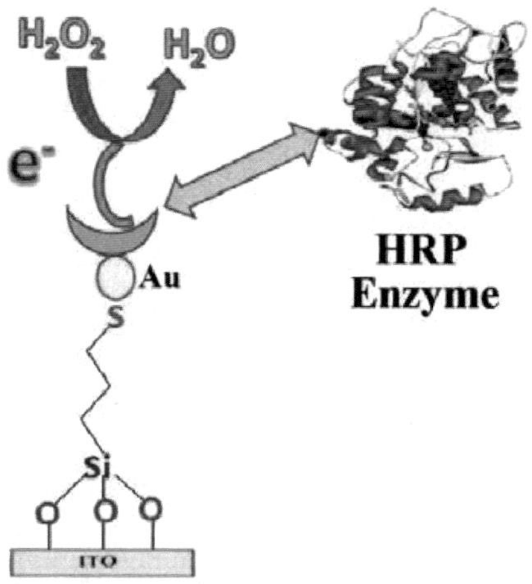

Scheme 1: Schematic representation of the proposed methodology for HRP immobilization on SAM modified ITO electrode.

2. Experimental section

2.1. Chemicals and materials

Horseradish peroxidase (HRP, RZ>3.0, A>250U/mg) enzyme was purchased from Sigma Aldrich. 3-mercaptopropyltrimethoxysilane (MPTMS), 80% (Across Organics) was used for monolayerformation on ITO electrodes. H_2O_2 (bio-chemica), KH_2PO_4 (HiMedia), NaOH (Merck) and other chemicals procured were of analytical grade. In addition potassium ferrocyanide, potassium ferricyanide, sodium chloride (all GR grade) were purchased from Merck. All these chemicals were used as received without any further purification. Millipore water having a resistivity of 18 MΩcm was used to prepare all the aqueous solutions used in this work. ITO sheets having a sheet resistance of $R_s = 4 - 8$□ were obtained from Delta Technologies Limited, Stillwater, MN, USA. This plate (25mm x 75mm x 0.7mm) is a single-side polished,

SiO_2 passivated float glass coated with ITO film of 200 – 500nm thickness. For other characterizations including electrochemical studies, this fairly big plate was cut into small pieces of pre-defined geometric area and was used as strips.

2.2. Preparation of characterization of MPTMS monolayer films on ITO electrodes

Before SAM formation, ITO strips were cleaned ultrasonically using acetone and water for 15min, respectively. After that it was immersed in an aqueous solution consisting of hydrogen peroxide, liquid ammonia, and water in the ratio of 1:3:5 for about 1h. Then these electrodes were rinsed several times with millipore water and immediately used for monolayer formation and subsequent analyses. We have used MPTMS (a short aliphatic chain with thiol terminal functionality) for SAM formation and it was prepared by immersing the pre-cleaned ITO electrodes into 1mM solution in toluene for about 14-15 hrs. After that the SAM of MPTMS coated ITO electrode was taken out; rinsed with excess of toluene and washed thoroughly with millipore water. Further, these electrodes were dried under a stream of nitrogen gas and immediately used for structural and electrochemical characterizations. XPS, AFM and contact angle measurements were used for the structural characterization of SAM modified electrodes.

Electron transfer (ET) behaviour and barrier property of these SAM modified ITO electrodes were evaluated by investigating the ET process of a redox probe using CV and EIS. A three-electrode electrochemical cell was used for electrochemical characterization. A platinum wire and saturated calomel electrode (SCE) were used as counter and reference electrodes respectively. Bare ITO, SAM modified ITO and HRP enzyme immobilized ITO electrodes were employed as working electrodes. Prior to use, the counter electrode was cleaned by dipping into 1:1 mixture of water and conc. HNO_3 for about 5min followed by a thorough cleaning with millipore water; and the reference electrode, SCE was rinsed thoroughly with millipore water. CV experiments were carried out at a potential scan rate of 50mV/s in an aqueous solution of 0.1M NaCl having 1mM potassium ferrocyanide as a redox probe within a potential range from -0.1V to 0.6V vs. SCE. Impedance measurements were carried out using an ac signal of 10mV amplitude at a formal potential of the redox probe (0.17V vs. SCE) using a wide frequency ranging from 100kHz to 100mHz. The electrolyte solution containing equal concentrations of both the oxidized and reduced forms namely, 1mM potassium ferrocyanide and 1mM potassium ferricyanide in 0.1M NaCl aqueous medium was employed for the study. In addition, the impedance data were useful in determining the kinetics parameters such as charge transfer resistance (R_{ct}), surface coverage (θ), pinholes – defects analysis and heterogeneous electron transfer rate constant (k_{ET}) associated with the monolayer film formation on ITO electrodes. All these experiments were performed at room temperature.

2.3. Immobilization of HRP enzyme on SAM modified ITO electrodes

SAM modified ITO electrodes were used as substrates for the immobilization of HRP enzyme. Basically we have adapted three methods. First method involves an electrostatic anchoring of HRP enzyme on SAM coated electrodes and in the other two methods gold nanoparticles (GNPs) were introduced to mediate the enzyme binding events. Second method consists of pre-concentration with copper followed by a galvanic replacement with gold and the other method involves the chemical interaction between GNPs and the thiolate (-SH) terminal functionality from the monolayer. Further these GNPs coated ITO electrodes were employed for HRP enzyme immobilization. During preconcentration step, initially SAM modified ITO electrode was dipped into 1mM aqueous solution of copper sulphate for 30mins. After that it was taken out; washed well with millipore water followed by potential cycling in a phosphate buffer solution with a pH of 7.4 within a potential ranges from -0.6V to 0.6V vs. SCE using CV. Finally, the Cu pre-concentrated ITO electrode was dipped into 1mM gold chloride ($AuCl_4$) solution for the galvanic replacement which is further confirmed by performing CV in aqueous phosphate buffer solution (pH = 7.4) between the potential range from -0.4V to 1.2V vs. SCE at a fixed scan rate of 50mV/s. In the case of third method, the monolayer coated ITO electrode was directly dipped into an aqueous solution of citrate stabilized GNPs for 30mins. CV and XPS were used to confirm the presence of GNPs. Finally these electrodes were dipped into 0.1M phosphate buffer solution containing HRP (0.1mg/2.5ml) for a specified time, which results in the anchoring of this enzyme on these chemically modified electrodes. After the immobilization, enzyme modified electrodes were washed well with phosphate buffer; thoroughly washed with millipore water and used for further characterizations and analysis.

2.4. Instrumentation

CV and EIS measurements were carried out using an electrochemical impedance analyzer, Model 6310, EG&G instruments obtained from Princeton Applied Research, USA and an AUTOLAB instrument, received from The Netherlands. CV experiments and their analysis of data were performed using *echem* software provided by EG&G. The potential ranges and scan rates used for the study were shown in the respective diagrams. The impedance data were fitted to an appropriate equivalent circuit model using the Autolab equipment and *Zsimpwin* software (from EG&G) developed on the basis of Boukamp's model. Based on this procedure several kinetics parameters mentioned above associated with the formation of monolayer were also determined. XPS measurements were carried out using Multilab 2000 model, Thermo Scientific, UK. A source of Al K_α having a binding energy of 1486.4eV was used for this study. Initially a full scan from -10 to 1100eV was performed and later on individual scans of 30-50eV having a step energy of 0.5eV were carried out in order to analyze the presence of specific elements and their corresponding oxidation states. AFM studies were carried out using a Pico Plus instrument made by Molecular Imaging (MI). The images were recorded in a tapping mode (ac mode) at a frequency of 175kHz with a cantilever of n_+-silicon type (PPP-NCL-50) obtained from Nanosensors, USA and these images were corrected for plane tilt using a software. The contact angle equipment was purchased from AST Products Inc., USA, having an automated model of VCA Optima XE. The measurement was carried out at room temperature on SAM modified substrates using water based on static sessile drop method.

3. Results and discussion

3.1. Characterization of SAM using XPS

ITO electrodes were chemically modified with a monolayer of MPTMS and the structural characterization of such a modified electrode was carried out using XPS, AFM and contact angle measurements. XPS is a surface chemical analysis technique in which the nature of adsorbing materials even up to the presence of a monolayer could easily be analyzed. We have investigated the formation of SAM of MPTMS on ITO using XPS and for comparison the bare ITO was also studied. Figure 1 shows the XPS spectra of S (2p), Si (2p), C (1s) and O (1s) regions of MPTMS monolayer on ITO surface. It can be seen that the formation of a peak at 163eV (2p) and at 154eV (2p) showed the presence of sulphur and silicon respectively. Further, the formation of peaks corresponding to carbon (1s) and oxygen (1s) could also be noted. Bare ITO showed a peak formation at 444.2eV ($3d_{5/2}$) and at 451.8eV ($3d_{3/2}$) corresponding to indium and at 486.3eV ($3d_{5/2}$) and 494.7eV ($3d_{3/2}$) corresponding to tin respectively, in addition to the usual carbon and oxygen peaks [19,20]. From XPS studies and analysis, it is clear that MPTMS molecules form a monolayer film on ITO electrodes.

3.2. AFM and contact angle measurements

Figure 2 displays a representative AFM image of SAM of MPTMS on ITO electrode. Formation of various domains (topography, 2A) of size ~15-20nm (surface profile, 2C) is clearly evident from these reproducible AFM images. We have measured a surface height of ~43-45nm in Z range from Fig. 2B (3D representation) and these domains extend vertically with the formation of pores in between them, suggesting the immobilization of MPTMS molecules onto ITO surface to form a monolayer film. Surface wettability in terms of hydrophilic and hydrophobic nature of the SAM modified ITO electrode was evaluated using contact angle measurements. We have used a static sessile drop method and the values reported were the average of contact angles measured at five different locations of the sample. Figure 2D shows the water static contact angle of monolayer film MPTMS on ITO electrodes. The equilibrium water contact angle of bare ITO (before modification) exhibits a strong dependence on the pre-treatment method adapted for cleaning the ITO surface. As received ITO surface provides a contact angle of 60±1°, which reflects a higher degree of hydrophobic nature of the substrate and upon pre-treatment by the method explained in experimental section 2.2, contact angle was found to decrease drastically to 18±2° implying a more hydrophilic nature. This change indicates a significant increase in the composition of hydroxyl groups on ITO surface as a result of this cleaning method. Similar observation was reported earlier in literature [21]. After the formation of MPTMS monolayer, contact angle value was found to increase significantly and was measured to be 76±1°.

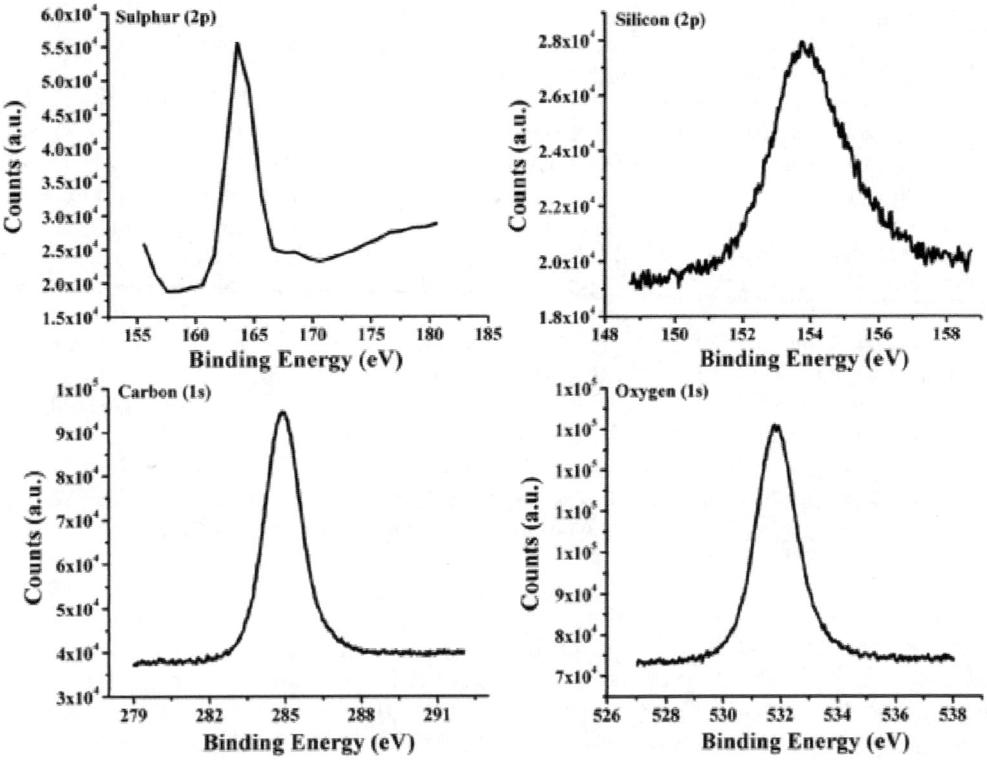

Figure 1: XPS spectra of MPTMS monolayer on ITO electrodes displaying S (2p), Si (2p), C (1s) and O (1s) regions.

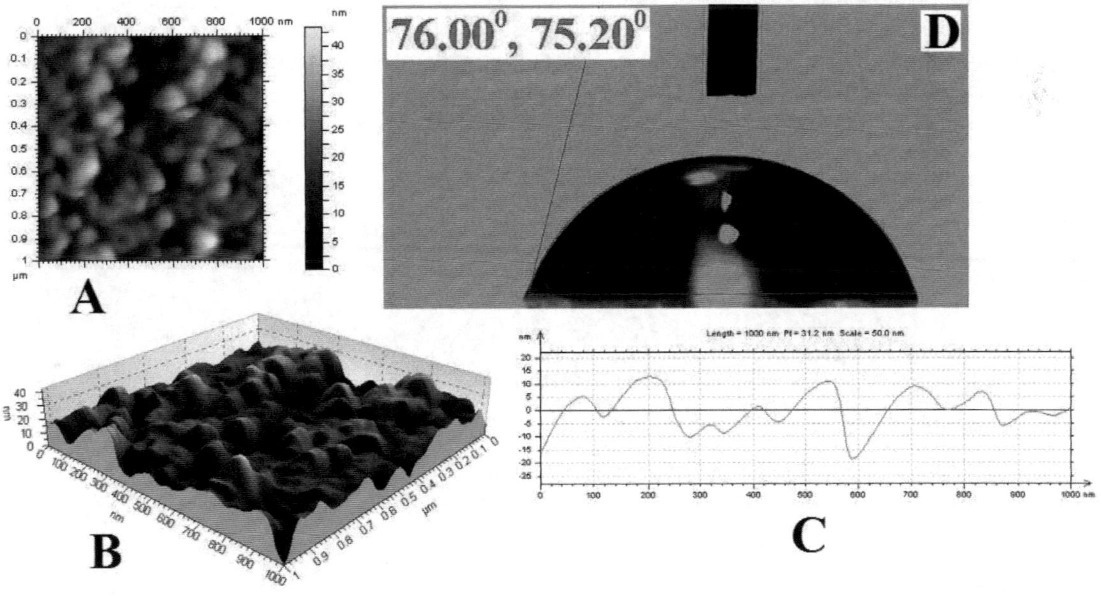

Figure 2: AFM [Topography (A), 3D representation (B) and surface profile (C)] and contact angle images of SAM of MPTMS on ITO electrode.

Journal of Physics: Conference Series **358** (2012) 012003 doi:10.1088/1742-6596/358/1/012003

3.3. Electrochemical characterization of SAM

CV and EIS are the most popular techniques in electrochemistry and these methods are demonstrated to be the most effective techniques to assess the quality of a monolayer film. Generally, the blocking behaviour of such monolayer films towards ET reaction of a redox probe is investigated using these techniques by employing potassium ferro/ferricyanide redox couple as a probe. Figure 3A shows the CVs of bare ITO and SAM of MPTMS coated ITO electrodes in 1mM potassium ferrocyanide with 0.1M NaCl as a supporting electrolyte at a potential scan rate of 50mV/s. It can be seen that bare ITO electrode (Fig. 3A[a]) exhibits a peak shaped CV for the redox reaction indicating that the ET reaction is under diffusion controlled and reversible. On the other hand, monolayer of MPTMS (Fig. 3A[b]) on ITO electrode does not show any distinct peak formation indicating that the ET reaction of a redox probe is completely suppressed and shows a perfect blocking behaviour with the formation of very less pinholes and defects. Interestingly, beyond a positive potential range of > 0.05V vs. SCE, the magnitude of redox current is found to be two orders of magnitude lower in the case of SAM of MPTMS when compared to bare ITO electrode. This clearly shows that the redox reaction is blocked on SAM modified electrodes due to the formation of highly ordered, well-packed, compact monolayer films with ultra low defect density.

Structural integrity and barrier property of this monolayer film were further evaluated using electrochemical impedance spectroscopic studies. Using this technique, a parameter, namely charge transfer resistance (R_{ct}) is determined, which provides the resistance offered by these monolayer films towards the ET process across the SAM modified electrode – electrolyte interface. In addition, the impedance data were also used to determine the surface coverage (θ) and other kinetic parameters of the monolayer coated electrodes. Figure 3B shows the impedance plots (Nyquist plots) of bare ITO and monolayer of MPTMS coated ITO electrodes in equal concentrations (1mM) of potassium ferrocyanide and potassium ferricyanide having 0.1M NaCl as a supporting electrolyte. The impedance measurements were carried out at a formal potential of $[Fe(CN)_6]^{3-/4-}$ redox couple, viz., 0.17V vs. SCE. It can be noted from these figures that bare ITO electrode (inset) shows a low frequency straight line with a very small semicircle at high frequency region indicating that the ET process of a redox couple is reversible and essentially diffusion controlled. On contrary, monolayer coated electrode exhibits a semicircle formation over the entire range of frequency used for the study indicating that the redox reaction is inhibited by displaying a perfect blocking behaviour. In such case the process of ET is considered to be under charge transfer control. Formation of a very large semicircle in the case of SAM of MPTMS on ITO electrodes confirms an excellent electrochemical blocking ability of these monolayer films against the diffusion of a redox probe molecule. Higher the charge transfer resistance, higher is the blocking behaviour. These results are in good agreement with our CV results discussed earlier.

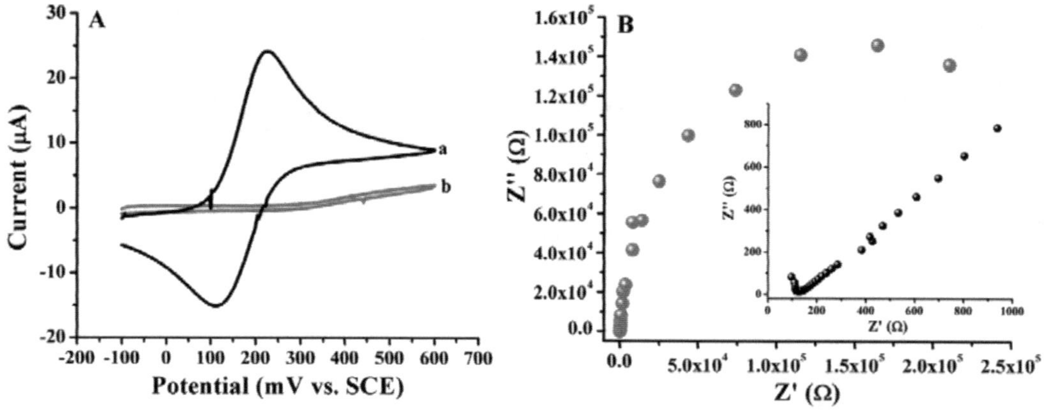

Figure 3: (A) CVs of bare ITO and SAM of MPTMS modified ITO electrodes in 1mM potassium ferrocyanide with 0.1M NaCl as a supporting electrolyte at a fixed scan rate of 50mV/s. (B) Impedance (Nyquist) plots in equal concentrations (1mM) of $K_4[Fe(CN)_6]$ and $K_3[Fe(CN)_6]$ aqueous solution containing 0.1M NaCl for the MPTMS monolayer on ITO. Inset shows the same plot for bare ITO electrode.

Journal of Physics: Conference Series **358** (2012) 012003 doi:10.1088/1742-6596/358/1/012003

By equivalent circuit fitting procedure using the impedance data we have determined R_{ct} values of 17.65Ω cm^2 and 5.44 x 10^{4}Ω cm^2 for bare ITO and SAM modified ITO electrodes respectively [22]. Basically the monolayer of MPTMS exhibited 4 orders of magnitude higher R_{ct} values implying a better blocking behaviour of SAM towards the redox reaction. Using this R_{ct} value, we have calculated the surface coverage of > 0.999 in the case of SAM of MPTMS suggesting the formation of wellpacked, highly dense monolayer films on ITO electrodes. From the R_{ct} values, one can also determine the heterogeneous electron transfer rate constant (k_{ET}) of $[Fe(CN)_6]^{3-/4-}$ redox couple on the unmodified bare ITO and SAM modified ITO electrodes using the following Eq. (1).

$$k_{ET} = RT / F^2 R_{ct} C \quad (1)$$

where R is the gas constant, T is the temperature, F is the Faraday's constant, C is the concentration of the redox probe and R_{ct} is the charge transfer resistance respectively. These values were calculated to be 0.1508 cm/s and 4.895 x 10^{-5} cm/s for the cases of bare ITO and MPTMS monolayer coated ITO electrodes [22]. It can be noticed that the monolayer displayed almost 4 orders of magnitude lower rate constant when compared to the rate constant of bare ITO electrode. This suggests that the SAM inhibits ET process by slowing down its kinetics due to the formation of a monolayer film on ITO electrode. The extent of sluggishness depends mainly on the quality and the chemical structure of monolayer films formed on ITO electrodes.

3.4. Immobilization of HRP enzyme

After SAM formation, GNPs were anchored onto the monolayer coated ITO electrodes. As mentioned earlier, two methods namely pre-concentration of Cu followed by a galvanic replacement with Au and functionalization of terminal thiol group with GNPs using chemical interactions. Both these methods result in the formation of ordered assemblies of GNPs on monolayer coated ITO electrodes and the presence of gold is confirmed using CV and XPS. In the case of CV carried out in an aqueous phosphate buffer solution at a constant scan rate of 50mV/s, appearance of redox peaks corresponding to Au reveals the anchoring of GNPs on SAM modified ITO electrodes. Further, the formation of Au 4f (5/2 and 7/2 spins) peaks at 84.4eV and 88eV during XPS studies confirms the presence of GNPs (Fig. 4A inset).

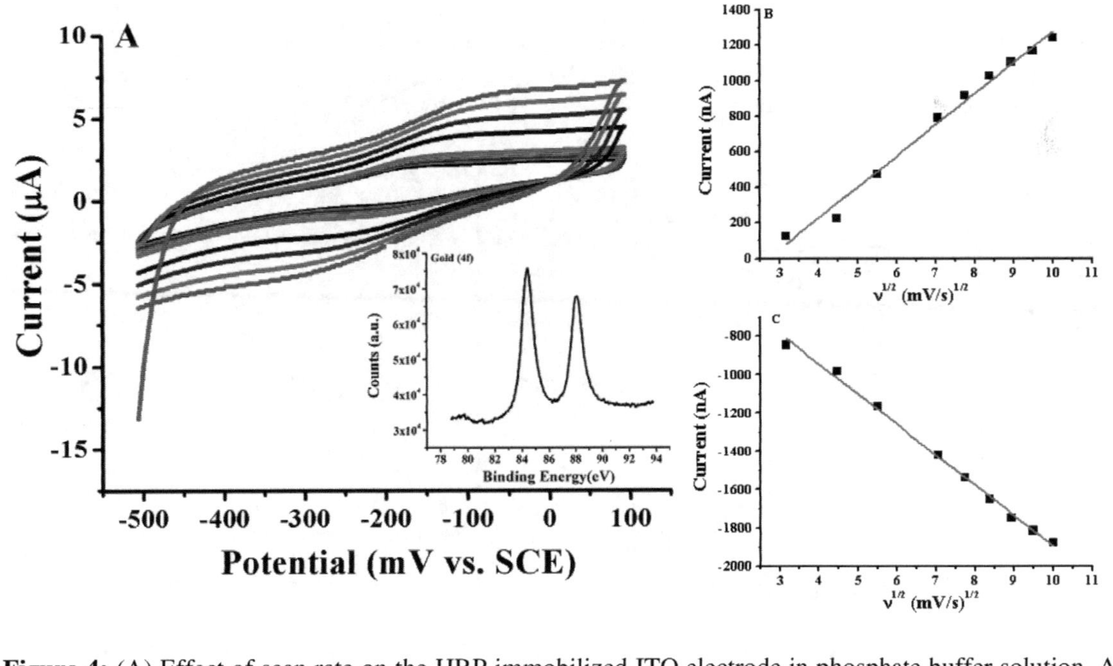

Figure 4: (A) Effect of scan rate on the HRP immobilized ITO electrode in phosphate buffer solution. A wide range of scan rate starting from 10mV/s to 100mV/s was used for the study. Inset shows the XPS spectrum of Au (4f) region. (B) and (C) show the plots of current (oxidation & reduction) vs. square root of scan rate.

HRP enzyme was immobilized on these GNPs coated electrodes by immersing them into a phosphate buffer solution consisting of HRP (0.1mg/2.5ml) for various amounts of time and evaluated their anchoring ability using CV. The redox properties and electron transfer ability of HRP enzyme were investigated with the aid of CV and these studies were carried out in an aqueous phosphate buffer solution possessing a pH of 7.4 at a fixed scan rate of 50mV/s within a potential ranging from -0.5V to 0.1V vs. SCE. The formation of redox peaks at -0.16V and -0.145V vs. SCE reveals the presence of redox proteins (due to Fe from HRP) confirming the immobilization of HRP enzyme on these modified ITO electrodes. Further the effect of scan rate on the redox property of HRP enzyme was studied by varying the scan rate from 10mV/s to 100mV/s and the corresponding CVs were displayed in figure 4A. The formation of redox peaks at all the scan rate suggests that the enzyme was immobilized on these modified electrodes and it is stable over a wide range of scan rate. Figures 4B and 4C show the plots of current corresponding to both the oxidation and reduction processes vs. square root of scan rate. It can be inferred from these plots that there is a linear relationship between the current and square root of scan rate suggesting that HRP enzyme is immobilized on these electrodes and its ET ability is under diffusion controlled. The fact that HRP immobilized ITO electrodes mediate direct ET across the interface implies that the redox sites present within the enzyme is accessible and these molecules structurally anchored on ITO electrodes in such a way that the redox protein is exposed outside. This sort of structural arrangement is very much essential to impart the sensing and catalytic applications so that the enzymatic kinetics can easily be monitored using potentiometry, amperometry and voltammetry. In addition XPS was also used to confirm the HRP immobilization and their results reveal the formation of Fe 2p peak at 719.6eV (Fig. 5A inset) corresponding to redox proteins of HRP enzyme.

3.5. Hydrogen peroxide (H_2O_2) sensing

Enzymatic reaction corresponding to HRP namely reduction of hydrogen peroxide (H_2O_2) to water was studied using these HRP immobilized ITO electrodes. These electrodes were initially tested towards H_2O_2 sensing in phosphate buffer solution and the corresponding studies were performed using CV. Figure 5A shows the CVs of HRP immobilized ITO electrodes in aqueous buffer solution towards H_2O_2 addition at a fixed scan rate of 50mV/s within the potential range between 0.1V and -0.5V vs. SCE. A wide range of concentration starting from 0.5mM to 5mM was used for the study. For comparison, a CV of HRP immobilized ITO electrode before H_2O_2 addition was also shown. It can be seen from these figures that the HRP anchored electrode displayed the redox peaks corresponding to the enzyme (confirmed by XPS – See inset) before the introduction of H_2O_2. On adding 0.5mM H_2O_2 to the buffer solution, a change in CV was

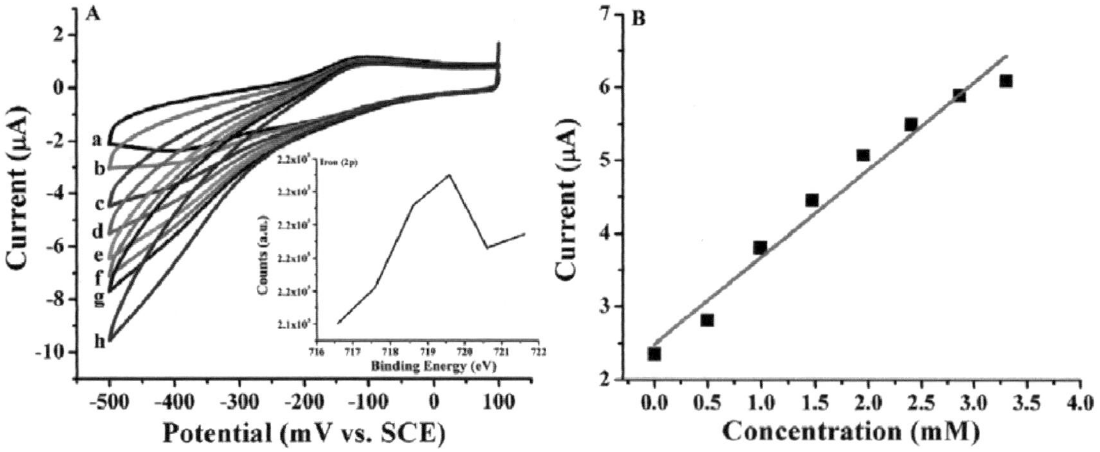

Figure 5: (A) CVs of HRP immobilized ITO electrode in a phosphate buffer aqueous solution towards H_2O_2 addition for various concentration ranging from (b) 0.5mM, (c) 1mM, (d) 1.5mM, (e) 2.0mM, (f) 2.5mM, (g) 3.0mM and (h) 3.5mM respectively. For comparison, the CV of HRP anchored electrode before H_2O_2 addition (a) was also shown. Inset shows the XPS spectra of Fe (2p) region of HRP enzyme immobilized ITO electrode. (B) A plot of current vs. concentration for H_2O_2 addition.

observed indicating a substantial increase in the magnitude of reduction current. Further increase in H_2O_2 concentration results in further increase in the reduction current values suggesting that HRP enzyme retains its activity and it helps in effectively reducing H_2O_2 to water. Figure 5B shows the corresponding plot of current vs. concentration of H_2O_2, from which sensitivity and detection limit values can be determined. In this case, we have determined a sensitivity of 1.2946μA/mM and the detection limit was estimated to be 2.858μM. These values suggest the better performance of this particular sensor when compared to some of the other works reported earlier. The higher sensitivity essentially arises from the introduction of GNPs which in turn helps in higher loading of HRP enzymes on chemically modified ITO electrodes. Further, we observed that either the monolayer of MPTMS coated or GNPs modified ITO electrodes do not contribute to sensing and reduction of H_2O_2 suggesting the effective role of HRP for both the sensing and catalytic applications.

4. Conclusions

Given the importance of bio-mimics, it is absolutely essential to understand the enzymatic reactions, structural arrangement and the associated electron transfer by immobilizing them onto the electrodes. In this work, we have successfully demonstrated a couple of strategies to immobilize and investigate enzymatic reaction of HRP using a simple process of chemical modification on ITO electrodes. Using this process, a sensor for H_2O_2 is fabricated and the performance is evaluated using CV and XPS.

Acknowledgement

Authors would like to acknowledge Department of Science and Technology (DST), India for providing funding towards this research work through Fast Track Scheme for Young Scientists under the project GAP 16/10. Useful discussions and help from Prof. V. Lakshminarayanan of RRI, Bangalore regarding AFM studies are acknowledged. Authors also thank Central Instrumentation Facility (CIF) of CSIR – CECRI for providing XPS facility for surface characterization.

References

1] Samanta D and Sakar A 2011 *Chem. Soc. Rev.* **40** 2567-2592, Gooding J J and Clampi S 2011 *Chem. Soc. Rev.* **40** 2704-2718
[2] Ronkainen N J, Halsall H B and Heineman W R 2010 *Chem. Soc. Rev.* **39** 1747-1763
[3] Nöll T and Nöll G 2011 *Chem. Soc. Rev.* **40** 3564-3576
[4] Wang J Lin Y and Chen L 1993 *Analyst* **118** 277-280
[5] Zhang J and Oyama M 2004 *Electrochim. Acta* **50** 85-90
[6] Matsubara C Kawamoto N and Takamura K 1992 *Analyst* **117** 1781-1784
[7] Pelossof G, Tel-Vered R, Elbaz J and Willner I 2010 *Anal. Chem.* **82** 4396-4402
[8] Jönsson G and Gorton L 1989 *Electroanal.* **1** 465-468
[9] Morales A, Céspedes F, Münoz J, Martínez-Fábregas E and Alegret S 1996 *Anal. Chim. Acta* **332** 131-138
[10] Wang L and Wang E 2004 *Electrochem. Commun.* **6** 225-229
[11] Qian L and Yang X 2006 *Talanta* **68** 721-727
[12] Tripathi V S, Kandimalla V B and Ju H 2006 *Biosens. Bioelectron.* **21** 1529-1535
[13] Liu Y, Geng T and Gao J 2008 *Microchim. Acta* **161** 241-248
[14] Bharathi S, Nogami M and Ikeda S 2001 *Langmuir* **17** 1-4
[15] Shipway A N, Lahav M and Willner I 2000 Adv. *Mater.* **12** 993-998
[16] Brown K R, Fox A P and Natan M J 1996 *J. Am. Chem. Soc.* **118** 1154-1157
[17] Zhang J D, Kambayashi M Oyama M 2005 *Electroanalysis* **17** 408-416
[18] Zhang J, Au K H, Zhu Z Q and O'Shea S 2004 *Optical Mater.* **26** 47-55
[19] Ganesh V, Maheswari D L and Berchmans S 2011 *Electrochim. Acta* **56** 1197-1207
[20] Ganesh V, Farzana S and Berchmans S 2011 *J. Power Sources* **196** 9890-9899
[21] Choi Y *et al* 2008 *Mol. Cryst. Liq. Cryst.* **492** 165.
[22] Muthurasu A and Ganesh V 2012 *J. Colloid Interface Sci.* **374** 241-249

Journal of Physics: Conference Series **358** (2012) 012004 doi:10.1088/1742-6596/358/1/012004

Culture experiments on conductive polymers

Mitsuyoshi Onoda[1]
Graduate School of Engineering, University of Hyogo at Himeji Shosha Campus, 2167 Shosha, Himeji, Hyogo 671-2280 Japan

E-mail: onoda@eng.u-hyogo.ac.jp or ss-syu02@eng.u-hyogo.ac.jp

Abstract. Fibroblast L929 and myoblast C2C12 cells of the mouse connective tissue origin were sown on the surface of conductive polymer films (polypyrrole, PPy and poly(3,4-ethylenedioxythiophene), PEDOT) in the cell culture medium, and the proliferative process of these cells was observed. Without changing the form, fibroblast L929 and myoblast C2C12 cells were observed to proliferate almost similarly to the cell which cultured on a dish on the market and to maintain compatibility. In other word, it has been understood these two kinds of conductive polymers used in this study, the PEDOT films maintain the secretion function of the cell cultured on the surface of these polymers. Therefore, the PPy- and the PEDOT-coated electrode suggested the possibility usable as a nerve stimulation electrode with biocompatibility, because these polymers were effective to culture the cell.

1. Introduction

The conventional common sense to be insulators that polymers don't conduct electricity is overturned, and polymers have alternating carbon-carbon single and double bonds turn into metal high conducting electricity by doping was found. For the development of a new field "Science and Technology of Synthetic Metals", the Nobel Prize in Chemistry at the 20th end of century was awarded jointly to Professor Alan J. Heeger from University of California, Santa Barbara, USA, Professor Alan G. MacDiarmid (death on February 7, 2007) from University of Pensylvanya, USA, and Professor of Emerit Hideki Shirakawa from Tsukuba University, Japan, "for the discovery and development of conductive polymers".

Conductive polymers have a structure that ranged long alternating carbon-carbon single and double bonds and extended highly conjugated system in their main chain. Because electrons contributing to double bond is easy to move along polymer main chain freely and have extended spatial and widely, the conductive polymers will show the specific properties that does not possess in conventional insulative polymers that polymer main chain consist of single bonds and only σ electrons exist. When thinking about conductive polymers from the viewpoint of electrical and electronic materials, though various kinds of conductive polymers are thought about whether are comprised of only carbon, hydrogen or also include the different kind atoms such as oxygen, nitrogen, sulfur etc., their nature depends on what kind of atoms is bonded how and forms molecules, in other words, being decided by each molecular structure. As one of the most characteristic effect that conductive polymers possess, doping effects are known well. By soaking or putting forcibly electron-accepting (acceptor) or electron-donating (donor) into the conductive polymer by means called the doping, the conductivity can be widely changed from insulator to semiconductor region, and then metal region and

[1] To whom any correspondence should be addressed.

accompanying this other properties including the magnetic and optical property change greatly, too. Such a doping effect is occurred by changing the electronic state greatly, because dopant delivers electrons with polymer main chain basically. As for such conductive macromolecule, the use as the ① metals, ② insulators and semiconductors, ③ insulator / metal transition phenomena, ④ reversible doping ability. and so on are possible [1]. By utilizing skillfully the property that conductive polymers possess, for example, practical use devices which application to electronic devices such as condensers, electroluminescent devices, photovoltaic cells, sensors, etc. including rechargeable batteries, and optoelectronic devices, are available began to be realized, and realization of the new electronic devices on organic materials mainly, in other words, the development and progress of the organic electronic technology are doing lively [2, 3].

On the other hand, in recent years, the research on the improvement and the reformation of medical equipment are actively examined with the advancement of remarkable medical technology. For instance, the nerve interface technology that directly inputs and outputs some information to man's nervous system is paid attention. If the signal of high accuracy can be inputted with a multichannel to the peripheral nervous system, the achievement of an impossible treatment method is expected in the prior art like a control of the external instrument by the movement instruction signal and an artificial sense by the external sense signal, etc. In order to achieve such a nerve interface technology, the biocompatibility that is appropriate for the burial of a long term can be possessed, meanwhile a steady signal can be inputted, and the development of the nerve stimulation electrode that has a number of channels enough for the control of the aimed device is indispensable. At present, though it is made with the metal electrode such as ITO (indium tin oxide) and platinum as a nerve stimulation electrode, the compatibility of the biological tissue and the metal electrode becomes a problem. As the measures, up to date the surface of the electrode was covered by having used collagen that was one of the matrices outside the cell (existing supermolecule structure besides the cell inorganism) and poly(lysine) that was the cell attachment factor (intercellular bonding) and compatibility was given [4]. However, it caused an increase in impedance by covering the surface of the electrode and decreased the efficiency of the input and output stimulus signals. Therefore, the development of the nerve stimulation electrode in which compatibility with the biological tissue is possessed, and the efficiency of the stimulus signal is not decreased is necessary, and it is extremely important in progressing the nerve interface technology.

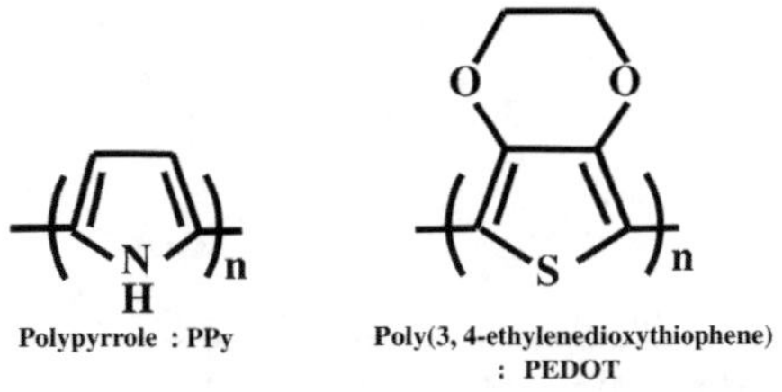

Figure 1. Examples of conductive polymer with good environmental stability.

In such a background, by thinking about the functional application of conductive polymer, the development to bioelectronics is also expected sufficiently. For example, thinking to use a conductive polymer as metal, namely an electrode, conductive polymers and the compatibility with the biocell become important. If such conductive polymers have a high biocompatibility compared with the metal, the functional application in biomedical engineering as a new artificial internal organmaterial is expected. A new progress and the progress of the nerve interface technology based on the development of the inside of the body burial type nerve stimulation electrode used from being conductive for the artificial sight, the artificial hearing or the artificial retina, etc. are thought. In developing the conductive polymers mentioned above as bioelectronic materials, the compatibility with the most basic and important conductive polymers and biocells have been reported in this article. To date, various conductive polymers have been synthesized, but poly(3,4-ethylenedioxythiophene),

(PEDOT), the molecular structure of which is shown in Fig. 1, is the most stable after doping and is one of the most attractive materials, together with polyaniline (PAn) and polypyrrole (PPy), because of its good environmental stability and relatively high electrical conductivity. Then, to develop nerve stimulation electrodes of the inside of the body burial type, PPy and/or PEDOT films of biocompatibility were mainly evaluated in the present study. For the biocompatibility of PPy and PEDOT films, the neuronal cell culture experiment was performed and evaluated. Fibroblastic cell (L929) removed from the mouse on the fabricated PPy and/or PEDOT films were cultured, and their survival and growth were observed. In addition, the culture experiment was also performed on myoblast (C2C12) removed from a mouse.

2. Experimental procedure

2.1. Fabrication of conductive polymers

In general, electrochemical polymerization technique is performed with the electrochemical polymerization equipment shown in Fig. 2. Though details of the electrolysis polymerization method are entrusted to technical books, the reaction mechanism of conductive polymer synthesis by electrochemical polymerization has not been clarified, because various factors, such as the composition of the electrolyte and the electrolytic conditions, interact very complexly with the electrode reaction. Therefore, the polymerization reaction condition differs for each individual conductive polymer and the optimum conditions are determined through trial and error. The mechanisms of the electrolysis polymerization reaction are not yet fully clarified under the present conditions, and qualitatively, both the parent electron substitution coupling reaction and the radical coupling reaction are accepted in general. At all events, in the electrolysis polymerization, he dynamic movement of the molecule, which takes part in the monomer, and the electrolytic ion formed by the transfer of an electron from the solvent are caused at the interface between the electrode and the electrolyte.

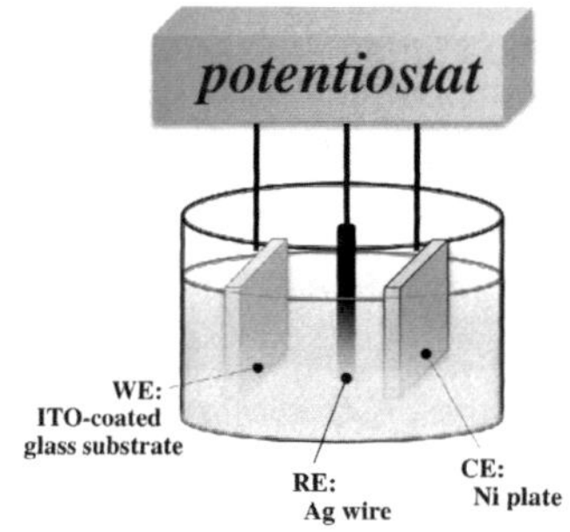

Figure 2. Schematic diagram of electrochemical polymerization equipment.

2.1.1. Fabrication of polypyrrole, PPy films[5]

The electrochemical polymerization was performed using an ITO (indium tin oxide) -coated conductive glass substrate with a working electrode, a nickel (Ni) plate as a counter electrode and a silver (Ag) wire as a reference electrode of the electrochemical cell as shown in Fig. 2. The composition of the polymerization liquid used for the electrolysis polymerization is 0.3 mol/l pyrrole (Tokyo Chemical Industry Co. Ltd.) as the monomer, and *tetra-n*-butylammonium tetrafluoroborate ($TBABF_4$, Tokyo Chemical Industry Co., Ltd. (TCI)), sodium *p*-toluenesulfonate (*p*-TSNa, TCI), *tetra-n*-butylammonium hydrogen sulphate ($TBAHSO_4$, TCI), *tetra*-methylammonium *p*-toluenesulphonate (TMA-*p*-TS, TCI), *tetra-n*-butylammonium bromide (TBABr, TCI), sodium *n*-dodecylbenzenesulfonate, LAS, TCI), *tetra-n*-butylammonium chloride (TBACl, TCI) *etc.* as the supporting electrolyte. The concentration of supporting electrolyte is changed within the range of 0.3–1.0 mol/l and then distilled water used for the solvent. The electrolysis polymerization was performed using a potentiostat (Hokuto Denko Inc., HSV-100), and the amount of passing electric charge at the polymerization potential 1.5 V (vs. Ag/Ag^+) was measured. Moreover, the thickness of the obtained PPy film was measured by the multiple interference method using a multiple interference microscope (Olympus, USPM-RU). The electrical conductivity of the obtained PPy film was about 30–50 S/cm.

2.1.2. Fabrication of poly(3,4-ethylenedioxythiophene), PEDOT films[5]

Jonas and Schrader [3] found that poly(3,4-ethylenedioxythiophene), PEDOT, with the molecular structure shown in Fig. 1, is extremely transparent and its stability is excellent under a typical environment. When the conductive polymer is actually applied, the environmental stability becomes extremely important. It is considered that doped PPy and polythiophene have excellent stability compared with that of positively charged polyacetylene, because of the stabilization effect of the positive charge with the nitrogen atom and the sulfur atom. On the other hand, because of the β position in PEDOT is oxidized easily and blocked by the ethylenedioxy groups, this material has high thermal stability and is excellent in high temperature and stable for a long time compared with PPy.

The electrolysis polymerization was performed by using ITO conductive glass substrate as a working electrode, Ni plate as the counter electrode and Ag wire as the reference electrode of the electrochemical polymerization device shown in Fig. 2. The composition of the polymerization liquid used for the electrolisis polymerization is 0.1 mol/l 3,4-ethylenedioxythiophene (Tokyo Kasei Inc., purity 98%) as the monomer and *tetra-n*-butylammonium tetrafluoroborate ($TBABF_4$, Tokyo Chemical Industry Co., Ltd.) as the supporting electrolyte. The concentration of supporting electrolyte is changed within the range of 0.3–1.0 mol/l. Then acetonitrile (Wako Pure Chemical Industries, Ltd., purity 99.5%) was used for the solvent. The electrolysis polymerization was performed using a potentiostat (Hokuto Denko Inc., HSV-100), and the amount of passing electric charge at the polymerization potential of 2.0 V vs. Ag/Ag+ was measured and the PEDOT film of about 400 nm in thickness was mainly obtained. The electrical conductivity of the obtained PEDOT film was about 120–200 S/cm.

2.2. Electrochemical properties of PEDOT films

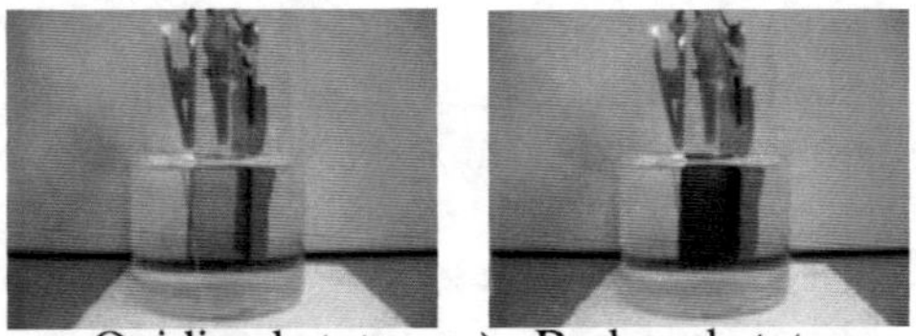

Figure 3. Color change of PEDOT film based on redox reaction.

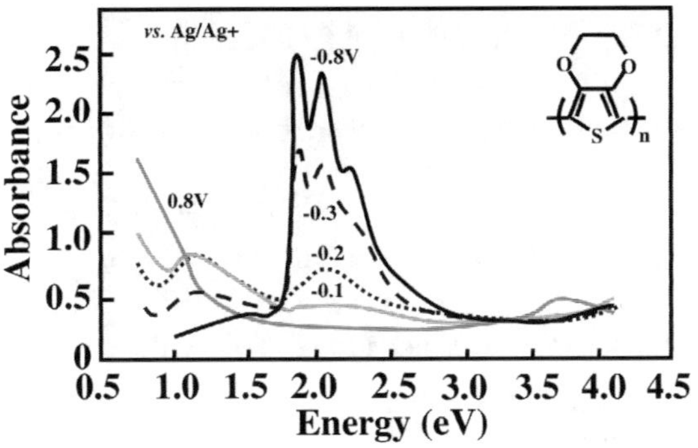

Figure 4. Optical absorption spectral change of PEDOT film taken during electrochemical doping.

Journal of Physics: Conference Series **358** (2012) 012004 doi:10.1088/1742-6596/358/1/012004

In general, it is well known electrical, optical or electrochemical properties of conductive polymers changed suddenly owing to the reversible doping. Fig. 3 shows the color change based on the redox reaction of PEDOT films. The PEDOT film is light blue or almost transparent in the doped state, and deep blue in the undoped state. And then, the color change is reversible in the PEDOT films.

On the other hand, Fig. 4 shows the change in the optical absorption spectrum of the PEDOT film with the doping potential. The films in the neutral and reduced states show strong absorption in the visible region and are deep blue in color. With increasing doping potential, the optical absorption in the visible region decreases and the film becomes light blue or transparent in the oxidized state from deep blue.

2.3. Subculture of cells

The subculture is to transfer the cells to other culture vessels. In this experiment, the subculture of fibroblast L929 and myoblast C2C12 was carried out respectively according to the following procedures. The dish used for the cell culture is a dish covered without collagen in fibroblast L929 culture and with collagen in myoblast C2C12. Fig. 5 shows the appearance of the cell culture.

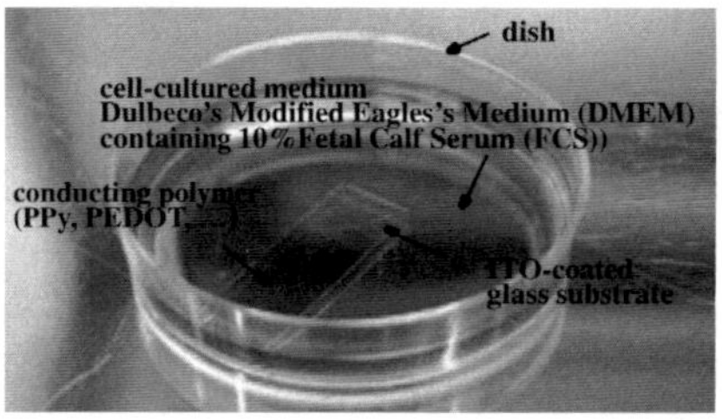

Figure 5. Appearance of cell culture.

2.3.1. The subculture of fibroblast L929 of mouse uniting organization origin

- The old nutrient medium is pulled out from the dish.
- The cell is washed adding PBS 5 ml (phosphate buffered saline).
- PBS is removed by sucking and because of peeling off of the cell that sticks at the bottom of the dish, 3 ml of trypsin solution is added and then the dish is left for about 5 min every time the CO_2 incubator reaches 37°C.
- Nutrientmedium Dulbeco's Modified Eagle's Medium (DMEM) containing serum 10% Fetal Calf Serum (FCS) content for amplification was added to the dish covered without collagen and the fibroblast L929was cultured

2.3.2. The subculture of fibroblast L929 of mouse uniting organization origin

- The old nutrient medium is pulled out from the dish.
- The cell is washed adding PBS 5 ml (phosphate buffered saline).
- PBS is removed by sucking and because of peeling off of the cell that sticks at the bottom of the dish, 3 ml of trypsin solution is added and then the dish is left for about 5 min every time the CO2 incubator reaches 37°C.
- This solution was putted in conical tube (BD Falcon) and the cells were precipitated by the centrifugal separator, and then trypsin solution was pulled out from the conical tube.
- Nutrientmedium Dulbeco's Modified Eagle's Medium (DMEM) containing serum 10% Fetal Calf Serum (FCS) content for amplification was added to the dish covered with collagen and the myoblast C2C12 was cultured.

In the culture experiment on the conductive polymer film, PPy and PEDOT were washed enough without peeling off from the ITO conductive glass substrate with distilled water, acetonitrile, and ethanol, in order to remove impurities adhered to the surface of the conductive polymer film and included impurities (oligomer etc. generated by unreactive monomer and the reactive process) during the electrolysis polymerization, and were dried. Conductive polymer coating ITO conductive glass substrate was put in the dish, and the subculture of the abovementioned was executed.

3. Experimental results and discussion

3.1. Culture experiment of fibroblastic cell L929

The fibroblastic cell is one of the cells that compose the mouse uniting organization, and its cytoplasm shows that nucleoli has clear oval nucleus and shows a base good nature. In this experiment, fibroblast L929 cells of the mouse uniting organization origin were used. Cultures were carried out in a humidified incubator with 5% CO2/95% air at 37 °C for several days. Fig. 6 shows the photograph of the cell observed with a headstand microscope (Nikon, TE2000-U) after 24 h sowing the fibroblast L929 cell to the dish. Fibroblast L929 cell bonding on the dish surface and progress to fusiform were observed.

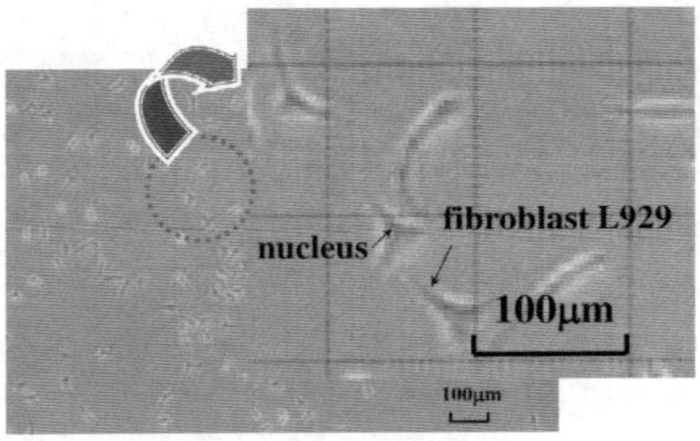

Figure 6. A macro photograph of fibroblast L929.

Fig. 7 shows the appearance of fibroblastic cell L929 culture. (a) on dish without collagen, (b) on PPy doped with TBABr and (c) on PEDOT doped with $TBABF_4$. From Fig. 7 (a), the cell proliferates on the dish surface and to all aspects of the dish in 96 h. On the other hand, from Fig. 7 (b) and (c), fibroblast L929 exhibited good adhesion on PPy doped with TBABr- and PEDOT doped with $TBABF_4$-coated ITO surfaces as well as the culture experiment that uses the dish. After 24 h, it is understood that the cell is bonded on conductive polymer films and proliferates. Ninety-six hours later, the cell proliferated to all aspects of the conductive polymer films, PPy and PEDOT, respectively. Though dopants that conductive polymer films possess differ from TBABr in the PPy film and $TBABF_4$ in the PEDOT film, it seems that the kind of dopants don't influence cell proliferation.

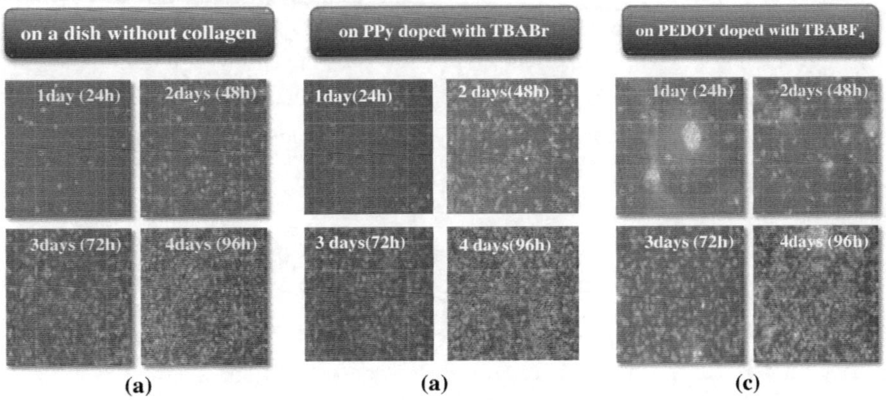

Figure 7. Appearance of fibroblastic cell L929 culture.

In microbiology, colony-forming unit (CFU) is an index of viable cell numbers and generally is given as CFU/ml (colony-forming units per milliliter) for liquids and CFU/g (colony-forming units per gram) for solids. In this study, the number of cells for each unit area was evaluated from this observation photographs in Fig. 7.

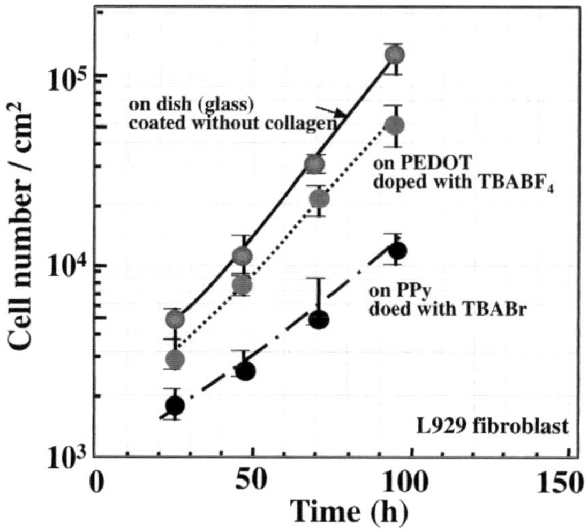

Figure 8. Relation between number of cells for each unit area of fibroblast L929 and culture days.

Fig. 8 shows the relation between number of cells for each unit area of fibroblast L929 and culture days. As shown in this figure, in the case of the culture on dish, though the content of the cell for each unit area is slightly a lot of, a remarkable difference between TBABr dope PPy and TBABF4 dope PEDOT cultured cells was not observed in four day culture. The doubling time of the number of cells evaluated from the inclination of this curve was about 23–28 h in the case of the culture on dish. It is reported that the doubling time becomes about 23 h on a collagen coating dish on the market [7]. On the other hand, the doubling time is evaluated for about 25 h on both PPy and PEDOT films obtained by electrolysis polymerization method, and it is considered that these conductive polymers are excellent culture substrates as well as the marketed culture dish.

3.2. Influence of concentration of supporting electrolyte on cell proliferation

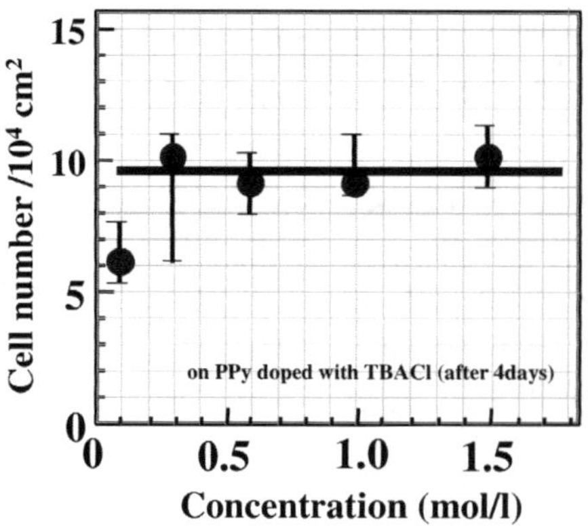

Figure 9. Relation between number of cells for each unit area and concentrations of supporting electrolyte.

Fig. 9 shows the relation between number of cells for each unit area of fibrolast L929 culture and the concentration of electrolyte, which evaluated from the observation result after 4 days culture on PPy doped with Cl^-. The composition of the polymerization liquid used for the electrochemical polymerization is 0.3 mol/l as the pyrrole as the monomer and TBACl as the supporting electrolyte and then acetonitrile (Wako Pure Chemical Industries, Ltd.) was used for the solvent. Though the number of culture cells is in the range of about $6\sim10\times10^{-4}$ pieces/cm^2 and has decreased in the low concentration, thinking about the influence of turbulence like the adhesion of the PPy film and conductive ITO glass substrate *etc.*, the proliferation of the cultured cell doesn't depend on the concentration of the supporting electrolyte.

3.3. Influence of the kind of supporting electrolyte on cell proliferation

In Section 3.1, it was described that the influence of the dopant that the conductive polymers possess seems to be not in the cell culture. Fig.10 shows the culture experimental results that used the PPy film to understand the influence by the kind of the dopant on the proliferation of the L929 fibroblastic cell.

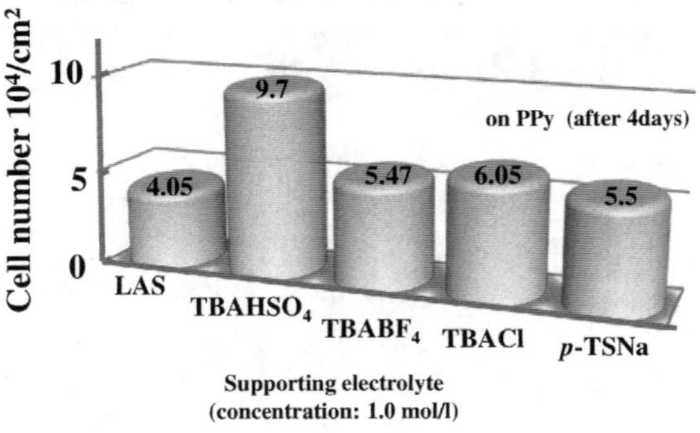

Figure 10. Relation between number of cells for each unit area and kinds of supporting electrolyte.

The supporting electrolyte used for the electrolysis polymerization is five kinds of LAS, $TBAHSO_4$, TBABr, TBACl , and *p*-TSNa. The concentration of supporting electrolyte is constant with 0.3 mol/l. The number of cells for each unit are after four days culture increases in the PPy film doped with $TBAHSO_4$, and the cell proliferation is apparently active. However, almost the same number of cells is obtained in the PPy film including other four kinds of dopants, and there is no difference between kinds of supporting electrolyte on the cell proliferation. However, the culture of the cell is greatly influenced also by adhesion of the ITO conductive glass used as a substrate and the PPy film. It is not possible to conclude from an actual experiment result alone. Many kinds of supporting electrolytes are used considering the problem of adhesion and a more detailed examination is necessary.

3.4. Myoblast C2C12 culture experiment

The myoblastic cell is the cell of a single nucleus that becomes an origin of the mouse myotube fiber, and when a lot of these cells fuse and form the syncytium, the myotube fibrocyte is formed. The cytoplasm of the myotube fiber is called sarcoplasm and is occupied in myofibril [8]. Myoblast C2C12 of the mouse skeletal muscle origin starts differentiation to a myotube fibrous cell by lowering serum density in culture fluid. Therefore, to differentiate the myoblast into myotube cells, culture fluid was changed for DMEM where a myoblast contained serum 7%HS (House Serum) for differentiation instructions in a subconfluent state (the state that multiplied to around 80% of the dish base) [9]. The culture fluid is changed every two days, and cell cultures were carried out in incubator under the environment of 100% of humidity with 5% CO_2/95% air at 37℃ for several days. The morphological change of cell according to cell culture period was observed by using a headstand microscope (Nikon, TE2000-U).

Journal of Physics: Conference Series **358** (2012) 012004 doi:10.1088/1742-6596/358/1/012004

(on PPy doped with *p*-TSNa)

(on PPy doped with TBACl)

Figure 11. Appearance of myoblast C2C12 culture.

Fig. 11 shows the appearance of the C2C12 myoblastic differentiation culture observed by using the PPy film polymerized from the electrolyte with the concentration 1 mol/l *p*-TSNa and TBACl as the supporting electrolytes. As is obvious from this figure, a differentiation instruction occurs after the subconfluent, and the myoblastic culture on the PPy film forms a myotube fibrous cell. In addition, for C2C12 myoblastic culture and a differentiation instruction, the influence of *p*-TS and the Cl dopant is not recognized.

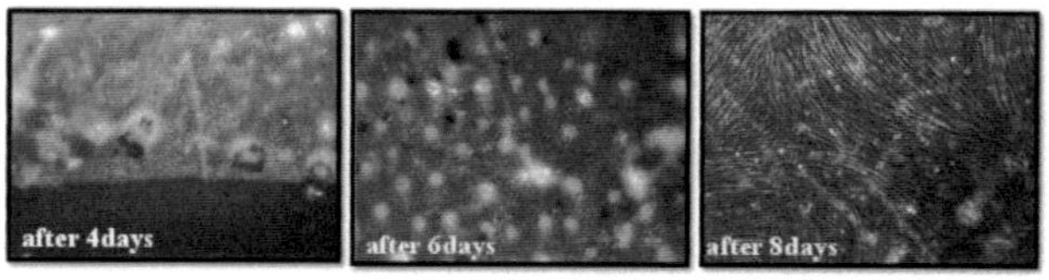

Figure 12. Appearance of myoblast C2C12 culture (on PEDOT doped with $TBABF_4$).

Fig.12 shows the appearance of the differentiation culture of the C2C12 myoblast observed by using PEDOT film fabricated from supporting electrolyte $TBABF_4$ of 0.3 mol/l in the concentration. As for the myoblastic culture on the PEDOT film, a differentiation instruction is not clear by the dirt of the lens from the appearance after for four and six days progress after subconfluent. However, from the appearance after 8 days progress, the occurrence of differentiation instruction and the formation of myotube fibrous cell are clear and BF_4 dopant does not influence the culture of the C2C12 myoblast and a differentiation instruction as well as PPy film. It is considered that these conductive polymers such as PPy and/or PEDOT films obtained by electrolysis polymerization method are excellent culture substrates and superior in biocompatibility as well as the marketed culture dish.

4. Summary

Fibroblast L929 and myoblast C2C12 cells of the mouse connective tissue origin were sown on the surface of conductive polymer films (PPy and/or PEDOT) in the cell culture medium, and the proliferative process of these cells was observed. Without changing the form, fibroblast L929 and myoblast C2C12 cells were observed to proliferate almost similarly to the cell which cultured on a dish on the market and to maintain compatibility. In addition, it was able to confirm that the myoblast C2C12 caused an induced differentiation on these conductive polymers films after the subconfluent and formed a myotube fibrous cell. In other words, two kinds of conductive polymers used in this study, PPy and PEDOT films maintained the secretion function of the cell cultured at the surface of these polymers. Therefore, the PPy- and the PEDOT-coated electrode suggested the possibility usable as a nerve stimulation electrode with biocompatibility because these polymers were effective to culture the cell.

References

[1] Yoshino K and Onoda M 1996 *Polymer Electronics* 266-367 (Japan-Corona) (in Japanese)
[2] Onoda M " Plastics through the electricity and electronics application" 2010 *Polyfile* **47**, No.559 22-27 (in Japanese)
[3] Onoda M "Expectation and problem to application as molecular electronics" 2011 *Industrial Materials* **59** No.8 42-46 (in Japanese)
[4] James C D Davis R Meyer M Turner A Turner S Withers G Kam L Banker G Craighead H, Isaacson M Turner J and Shain W "Aligned Microcontact Printing of Micrometer-Scale Poly-Llysine Structures for Controlled Growth of Cultured Neurous on Planar Microelectrode Arrays" 2000 *IEEE Transactions on Biomedical Engineering*, **47** No. 1 17-21
[5] Abe Y Mathur P C Bhatnagar P Tada K and Onoda M "New fabrication technique of conductive polymer/insulating polymer composite films" 2008 *IEE Journal of Transactions on Fundamentals and Materials* **128** No.12 703-709 (in Japanese)
[6] Jonas F and Schrader L "Conductive Modification of Polymers with Polypyrrole and Polythiophene" 1991 *Synthetic Metals* **41-43** 831-836
[7] Yamauchi K, Maniwa M Mori T "Cultivation of fibroblast cells on keratin-coated substrata" 1998 *Journal of Biomaterials Science, Polymer Edition* **9** No.3 259-270
[8] Gerard J Tortora Bryan Derrickson 2010 "*Principles of Anatomy and Physiology* (12th Editions). Translated by Kuwaki T Kurosawa M Takahashi K and Hosotani Y 3rd Editions 302-308 (Japan-Maruzen) (in Japanese)
[9] Kubo Y "Comparison of initial stages of muscle differentiation in rat and mouse myoblastic and mouse mesodermal stem cell lines" 1991 *Journal of Physiology*, **442** No.1 743-759

Transmitter function of synapse-structure system using conducting polymer

Masaharu Fujii[1], **Yuka Machiya and Haruo Ihori**
Graduate School of Engineering, Ehime University,
3 Bunkyocho, Matsuyama, Ehime 790-8577, Japan

E-mail: fujii.masaharu.my@ehime-u.ac.jp

Abstract. Conducting polymers with neuron-like pattern has been polymerized by controlling polymerization conditions. These conducting polymers have been connected each other to prepare network. If the synapse function can be added to the network, artificial neural network is prepared by conducting polymer. In this paper, we consider the transmitter function using synapse-structure conducting polymer. It consists of three parts: primary circuit as presynaptic terminal, space as synaptic cleft and secondary circuit as postsynaptic structure. Dopant in conducting polymer works as neurotransmitter. Migration as well as diffusion is also considered for dopant ion to transit the space/cleft. When signals from the primary circuit came at the end of the primary circuit in electrolyte solution, the current in the secondary circuit increased because the released dopant ion transited the cleft and entered another conducting polymer. When two primary circuits was used, the current in the secondary circuit increased higher than one primary circuit. This means the synapse-structured conducting polymer system can be use as logical circuit.

1. Introduction

Conducting polymers with neuron-like pattern have been polymerized by controlling the polymerization conditions. The neuron-like conducting polymer can be connected to prepare a network. The neuron-like or neural network device can be fabricated if three functions are added to the network of conducting polymer: (i) multilayer, (ii) threshold function, and (iii) controlled weight of network-path. Two methods have been suggested to fabricate the device neuron-like device: (1) Chemical and (2) electrical method. Furthermore, logical circuits (AND/OR circuits) have been fabricated by these networks. Meanwhile conducting polymers have been used in bioelectronics [1-].

Neural network in computer field, Perceptron, was investigated by considering neuron work, especially synapse function. Three functions ((i)-(iii)) described above are obtained by modeling synapse to make a system of brain. If the function at the synapse connection is realized by synapse structure; (a)presynaptic terminal, (b)synaptic cleft and (c)postsynaptic structure, threshold and weight characteristics can be easily realized.

In this paper, we consider the transmitter function using synapse-structure conducting polymer.

2. Synapse-structured using conducting polymer

[1] To whom any correspondence should be addressed.

The synaptic junction consists of three parts. When signals arrive at the presynaptic terminal, neurontransmitters are released from presynaptic terminal to synaptic clef and interact with receptor molecules in the postsynaptic membrance. They finally exert an effect on the membrance potential of the neuron.

The synapse structure using conducting polymer is also the same structure: primary conducting polymer as presynaptic terminal, secondary conducting polymer as postsynaptic structure and small gap as synaptic clef. When the pulses arrive at the primary conducting polymer, dopant ions in the primary conducting polymer are released in the gap. The dopant ions released diffuse toward the secondary conducting polymer. The concentration of dopant ions increases near the secondary conducting polymer and the dopant ions penetrate the secondary conducting polymer. Then the conductivity of the secondary conducting polymer increases. Current in secondary circuit with the secondary conducting polymer increases.

3. Experimental

Figure 1 shows the experimental configuration. The primary circuit consisted of a pulse voltage source, conducting polymer with dopants and a counter electrode of Ni. The input is a series of 5 [V] rectangle pulses whose width and period are 0.5 [sec] and 1.0 [sec] receptively. The secondary circuit consists of a source voltage and a film of conducting polymer undoped. The conducting polymer undoped was set between the input electrode, which is the conducting polymer doped, and the counter electrode. When the pulse voltage is applied, the ions penetrates the input-side of conducting polymer due to the potential slope, i.e. the ions drift into the input-side of conducting polymer. Thus the current in the secondary circuit increases.

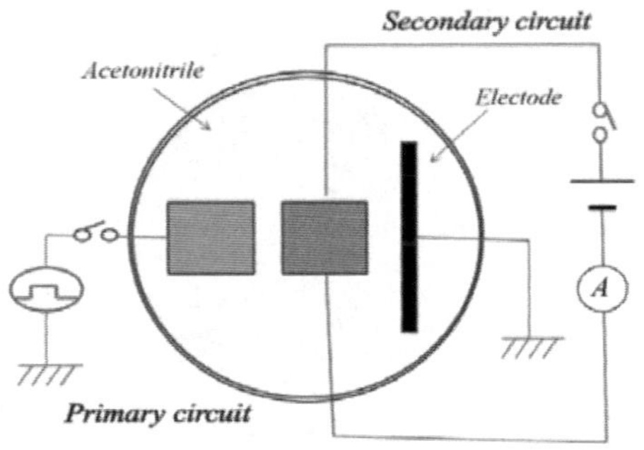

Figure1. Experimental configuration.

4. Results and discussion

Figure 2 shows the secondary circuit current under three trials. The current increased after each trial of adding a series of pulsed. The dopant ions drifted from the input electrode of conducting polymer doped to the secondary conducting polymer undoped. The cross potential system assisted the transit of the dopant/transmitter. In the system the number of pulses control amount of dopant, which change the conductivity of the secondary conducting polymer. That should work as the weight and threshold of neural network.

To investigate the effect of the number of pulse effect on secondary conducting polymer, two Ni electrodes were used instead of conducting polymer electrodes and conducting polymer in secondary circuit was polypyrrole prepared by electrolyte of $LiBF_{4}$/ sodium *p*-toluenesulfonate. Figure 3 shows the experimental configuration with two primary circuits(x1 and x2). Figure 4 shows the addition of two input of x1 and x2. One input consisted of 15 pulses. When one input was on, the output of current was almost same. When double inputs were added(x1 + x2), the output current was higher than one input. It is hard to use this system for additional counter because of nonlinearity of conductivity but it works as AND logical circuit if the threshold is set at 0.073mA in this case.

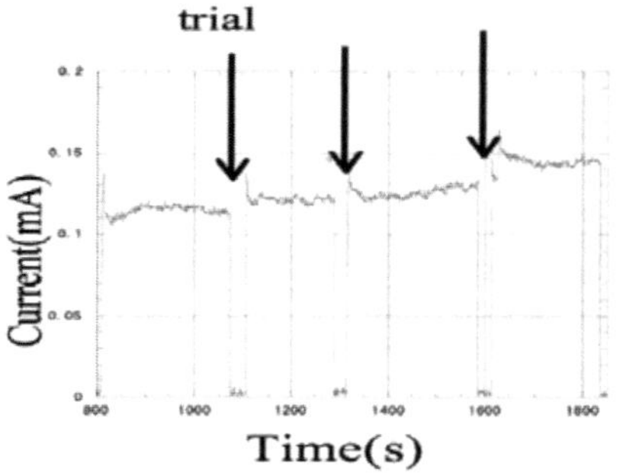

Figure 2. Current in secondary circuit.

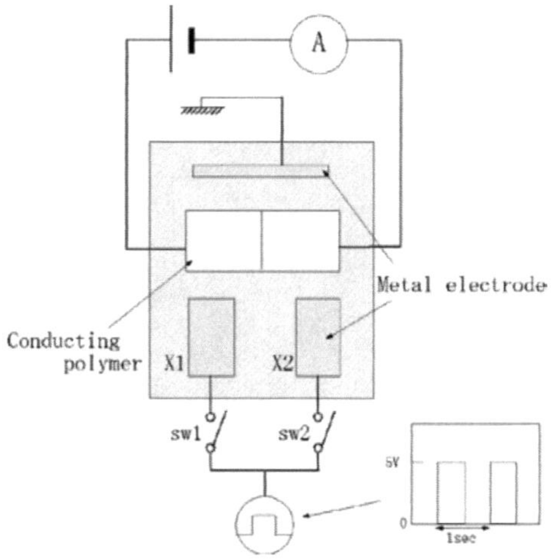

Figure 3. Two-inputs system.

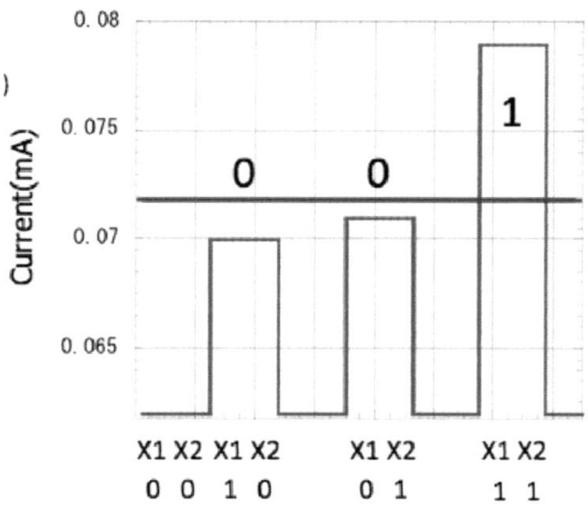

Figure 4. Output current of two inputs x1 and x2.

5. Conclusion

The synapse-structured conducting polymer has been investigated for transmitter action of synapse. The structure using conducting polymer consisted of three parts; the primary circuit, the gap and the secondary circuit corresponding to presynaptic terminal, synaptic cleft and postsynaptic structure, respectively. The current through the secondary conducting polymer increased after each trial. The dopant ion worked as transmitter. The gradient of potential introduced has been used to accelerate the ions.

Acknowledgments

This work was supported by KAKNHI(Grant-in-Aid for Scientific Research (C)22560298).

References

[1] Swann M J, Bloor D, Haruyama T and Aizawa M 1997 *Biosensors and Bioelectronics* **12** 1169
[2] Cuia X, Wiler J, Dzaman M, Altschuler R A and Martin D C 2003 *Biomaterials* **24** 777
[3] Georgea P M, Lyckman A W, LaVan D A, Hegde A, Leung Y, Avasare R, Testa C, Alexander P M, Langer R and Sur M 2005 *Biomaterials* **26** 3511

Electrochemical Urea Biosensor Based on Sol-gel Derived Nanostructured Cerium Oxide

Anees A Ansari[1]*, Md. Azahar[2] and B. D. Malhotra[3*]
[1]King Abdullah Institute for Nanotechnology, King Saud University, Riyadh, 11451 Kingdom of Saudi Arabia
[2]Department of Science & Technology Centre on Biomolecular Electronics,National Physical Laboratory, Dr. K S Krishnan Marg, New Delhi-110012, India
[3]Department of Biotechnology, Delhi Technological University,Shabad Daulatpur, Main Bawana Road, Delhi 11042,India

E-mail: aneesaansari@gmail.com ; bansi.malhotra@gmail.com

Abstract. Urease (Urs) and glutamate dehydrogenase (GLDH) have been co-immobilized onto a nanostructured–cerium oxide (Nano-CeO_2) film deposited onto a indium-tin-oxide (ITO) coated glass substrate by dip-coating via sol-gel process for urea detection. This nanostructured film has characterized using X-ray diffraction (XRD), Fourier transform infrared (FTIR), Scanning electron microscope (SEM) and electrochemical techniques, respectively. The particle size of the Nano–CeO_2 film has been found to be 23 nm. Electrochemcial response (CV) studies show that Ur-GLDH/Nano-CeO_2/ITO bioelectrode is found to be sensitive in the 10–80 mg/dL urea concentration range and can detect urea concentration upto 0.1 mg/dL level. The value of Michaelis–Menten constant (K_m) estimated using Lineweaver–Burke plot found as 6.09 mg/dL indicates enhancement in the affinity and/or activity of enzyme attached to their nanobiocomposite. This bioelectrode retained 95% of enzyme activity after 6 months at 4°C.

1. Introduction

Novel analytical devices based on nanostructured metal oxides are known to be cost-effective and highly sensitive due to the large surface-to-volume ratio, show excellent selectivity and their optical and electrical properties arising from electron and phonon confinement. These nanostructured materials can be used as sensing biomolecules[1-18]. Besides these nanostructured metal oxides have unique properties such as non-toxicity, biocompatiblity, high chemical and thermal stability, high isoelectric point (IEP ~9.2) and negligible swelling in both aqueous and non-aqueous solvents[1-6]. A large number of nanomaterials such as carbon nanotubes[7], conducting polymers[8], grapheme[9], nanowires[10], zinc oxide nanoparticles and chitosan composite[11] film and nanoporous materials have been used for fabrication of various biosensors. Many nanostructured metal oxide such as zirconium oxide (ZrO_2), tin oxide (SnO_2), cerium oxide (CeO_2) and zinc oxide (ZnO) have been utilized for immobilization of proteins, enzymes and antigens for accelerated electron transfer between active sites of protein and electrode[12-13]. On the other hand, MnO_2 and TiO_2 having low IEP values are suitable for the immobilization of high IEP proteins.

* To whom any correspondence should be addressed.

Recent times have been significant advances towards the production of nanocrystalline metal oxides using a variety of physical and chemical processes, such as sputtering, sol–gel, hydrothermal, spray pyrolysis etc. Among the various methods sol-gel technique has been considered as very attractive for the preparation of metal oxide particles due to easy preparation under ambient conditions, tunable porosity, high thermal stability, chemical inertness and negligible swelling in aqueous and non-aqueous solutions[14-15]. Besides this, due to inherent low temperature process, the sol-gel technology provides an attractive method for the immobilization of heat-sensitive biomolecules such as enzymes, proteins and antibodies[16]. The positively charged sol–gel cerium oxide not only provides a friendly environment for the enzyme to retain its activity but also encapsulate the negatively charged electron transfer mediator within the sol-gel film via electrostatic interaction[17]. The sensitivity and stability of the biosensors can perhaps be improved by controlling charge, porosity and electronical conductivity of a given sol-gel film. These attractive features have led to an increased research towards developing the electrochemical sensors and biosensors[18].

The estimation of urea is of clinical interest since decreased urea concentration (normal range is 15-40 mg/dl) causes hepatic failure, nephritic syndrome and cachexia and increased urea level in blood and urine causes renal failure, urinary tract obstruction, dehydration, shock, burns and gastrointestinal bleeding[19-21]. Therefore, development of rapid, cheap and reliable methods for urea determination is important. Among the numerous analytical methods reported in literature, enzymatic methods offer an improvement in terms of specificity, response time, miniature size, reproducibility and selectivity. Saha et. al. fabricated cerium oxide (CeO_2) film deposited onto platinum (Pt) electrode using pulsed laser deposition (PLD) has been utilized for immobilization of glucose oxidase (GOx)[22]. Feng *et. al.* have prepared a nanoporous CeO_2/chitosan composite matrix for the immobilization of single-stranded DNA probes for the detection of cancer genes[23]. Khan et. al. ZnO-chitosan nanocomposite has recently been used for application to urea and cholesterol biosensors[11].

We have used a facile sol-gel chemical route to prepare Nano-CeO_2 film onto indium-tin-oxide glass substrate (ITO) to immobilize a mixture of ureas and GLDH enzymes for application to urea detection using electrochemical spectroscopy technique.

2. Experimental procedure

2.1. Chemicals and reagents

Urease (Urs), glutamate dehydrogenase (GLDH), nicotinamide adenine dinucleotide (NADH) α-keto glutarate (α-KG), Na_2HPO_4 and NaH_2PO_4 were procured from Sigma Aldrich (USA). Ammonium ceric nitrate, urea, NaCl, NH_4OH and HNO_3 reagents were procured from Merck India Ltd, Mumbai, India. All these chemicals were of analytical grade and used without further purification. Indium-tin-oxide (ITO) coated glass plates were obtained from Balzers, UK. The deionized water obtained from Millipore water purification system (Milli Q 10 TS) was used for preparation of solutions and buffers.

2.2. Fabrication of sol-gel derived nano-structured CeO_2 film/electrode

1 gm of ceric ammonium nitrate [$(NH_4)_2Ce(NO_3)_6$] was dissolved in deionised water (20 ml). Then 5 ml (1 M) solution of ammonium hydroxide (NH_4OH) was added drop wise in this solution with constant stirring for 4 h at 25 °C to maintain pH 10. A pale yellow precipitate of $Ce(OH)_4$ thus obtained was washed several times with deionized water until a neutral pH was achieved[13]. Subsequently, dilute HNO_3 (1M) was added to the precipitate at 60 °C to obtain a solution of pH 1. A transparent yellow solution thus obtained was further concentrated by heating. The resulting sol was used to fabricate thin film on ITO coated glass plate via dip coating technique and was allowed to dry at 500 °C.

2.3. Electrode modification and immobilization Urease-GLDH

10 μL of 1:1 molar mixture of urease and GLDH (1.0 mg/mL, in PBS, 50 mM, pH 7.0) was immobilized onto sol-gel derived Nano-CeO_2 electrode by physisorption method. The biochemical reaction using mixed enzyme system is shown in Eq.1. Prior to being used, Urs-GLDH/Nano-CeO_2/ITO bioelectrode was allowed to dry overnight under desiccated conditions and then washed with phosphate buffer saline (PBS, 50 mM, pH 7.0, 0.9 % NaCl) to remove any unabsorbed enzymes (Urs-GLDH) and stored in a desiccator at 25°C when not in use.

$$\text{Urea} \xrightarrow{\text{urease}} 2NH_4^+ + HCO_3^- \quad \text{(Eq. 1)}$$
$$2NH_4^+ + \alpha\text{- ketoglutarate} + \text{NADH} \xrightarrow{\text{GLDH}} \text{2- Glutamate} + \text{NAD}^+ \quad \text{(Eq. 2)}$$
$$\text{NADH} \longrightarrow \text{NAD}^+ + 2e^- \quad \text{(Eq. 3)}$$

2.4. Characterization of sol-gel derived Nano-CeO_2/ITO electrode and Urs-GLDH/Nano-CeO_2/ITO bioelectrode

X-ray diffraction (XRD, Cu Ka radiation (Rigaku)) study was performed to identify the crystal structure of sol-gel derived Nano-CeO_2 film. Scanning electron microscopy (SEM, LEO-440) studies were conducted to examine the surface morphology. FTIR spectra of sol-gel derived Nano-CeO_2 films were recorded using FTIR (Perkin-Elmer) spectrophotometer to investigate binding of ureas enzyme onto sol-gel derived Nano-CeO_2/ITO film. Electrochemical measurements were conducted on an Autolab Potentiostat/Galvanostat (Eco Chemie, Netherlands) using a three-electrode cell containing ITO act as a working electrode, Ag/AgCl as reference electrode and platinum (Pt) wire serve as a counter electrode in phosphate buffer saline (PBS, 50 mM, pH 7.0, 0.9% NaCl) containing 5 mM $[Fe(CN)_6]^{3-/4-}$.

3. Results and discussion

3.1. Characterization of nano- CeO_2 film

The XRD diffraction pattern (Figure 1) shows a crystallographic phase present in the sol–gel derived CeO_2 film deposited on the glass substrate via dip-coating technique. A high degree of preferential orientation is evident, giving rise to spectra resembling a single crystal diffraction pattern. XRD pattern of the deposited film reveals all reflection planes of CeO_2, corresponding to the (111), (200), (220) and (311) crystallographic plane cubic fluorite structure (space group Fm3m (225)) as identified using the standard data JCPDS Card No. 34-0394. The intensities and positions of the peaks are in perfect agreement with the literature values. No peak of any other phase is detected indicating the high purity of the material. The broadening of the reflections in the diffractogram distinctly indicates the formation of nano-dimensional crystals. The average crystallite size of CeO_2 nanoparticles was estimated by analysis of the broadening of (111) and (220) reflections via Scherrer formula found to be 10-15 nm.

3.2. Characterization of nano- CeO_2 film

The infrared absorption spectra of Nano-CeO_2/ITO (a) and Urs-GLDH/Nano-CeO_2/ITO (b) shows the characteristic absorption bands of the deposited films (Figure 2). A diffused band at 3404 cm^{-1} and two weak bands at 1414 and 1010 cm^{-1}, corresponding to the stretching and bending vibrations of the hydroxyl (O–H) groups, respectively, are observed. The appearance of these bands suggests the adsorption of moisture on the surface of nanostructured film. The FTIR spectrum shows a sharp and intense band at around 426 cm^{-1} assigned to the Ce-O stretching band and indicates the deposition of CeO_2 film on the ITO surface. These observed bands are shifted and some new bands are appearing at 1740, 1380 and 1050 cm^{-1} corresponding to the stretching and bending vibrations of the amide, the functional groups. This indicates that the Urs-GLDH mixed enzymes solution has been successfully immobilized on the Nano-CeO_2 matrix. A sharp band at 450 cm^{-1} is observed due to the electrostatic interaction and hydrogen bonding of the Nano-CeO_2 with the enzymes (Urs-GLDH).

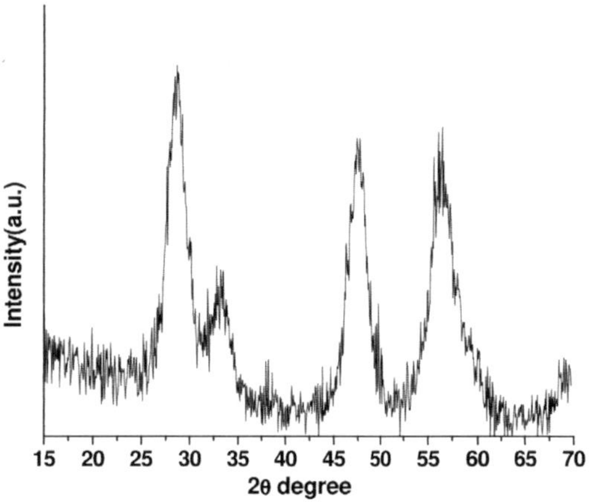

Figure 1. XRD pattern of sol-gel derived CeO_2 film

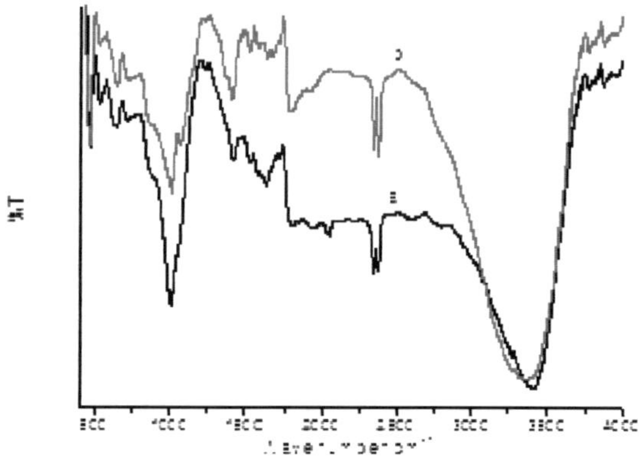

Figure 2. FTIR spectra of (a) sol-gel derived Nano-CeO_2/ITO electrode and (b) Urs-GLDH/Nano-CeO_2/ITO bioelectrode

3.3. Scanning electron microscopy studies

Figure 3 shows results of the SEM studies carried out on Nano-CeO_2/ITO film and of Urs-GLDH/Nano-CeO_2/ITO bioelectrode, respectively. The SEM micrograph shows that the film surface is free of cracks, uniform in thickness porous in nature and composed of very small fine grains that are homogenously distributed (image 3a). The adsorption or immobilization of urease enzyme onto the surface of sol-gel derived Nano-CeO_2/ITO electrode is dependent on the morphology of the substrate. It can be seen that after immobilization of Urs-GLDH onto Nano-CeO_2/ITO electrode, displays flower like structure with high porosity, inter-connectivity and significant free volume favoring the effective immobilization of a large number of enzymes (image 3b).

3.4. Rlectrochemical impedance spectroscopy

EIS measurements carried out on Nano-CeO_2/ITO electrode and Urs-GLDH/Nano–CeO_2/ITO bioelectrode at frequency range 0.0–5×10^3 Ω. It is an effective and powerful tool for characterizing the interfacial features of surface-modified electrodes. The modified electrode impedance can be presented as the sum of the real (Z'), and imaginary ($-Z''$) components that originate mainly from the resistance and capacitance of the cell, respectively. The general electronic equivalent circuit (Randles and Ershler model), includes the ohmic resistance of the electrolyte solution (R_s), the Warburg impedance (D), resulting from the diffusion of ions from the bulk electrolyte to the electrode interface. The double layer capacitance (C_{dl}) and charge-transfer resistance (R_{ct}) exists, if a redox probe is present in the electrolyte solution, where R_s and D denote bulk properties of the electrolyte solution and diffusion features of the redox probe in solution, respectively. The other two components C_{dl} and R_{ct}, depend on the dielectric and insulating features at the electrode/electrolyte interface. Figure 4 shows the Faradaic impedance spectra, presented as Nyquist plots obtained from real (Z'') and imaginary ($-Z''$) of blank ITO (curve a), Nano-CeO_2/ITO electrode (curve b) and Urs-GLDH/Nano-CeO_2/ITO (curve c) bioelectrode. The values of R_{ct} derived from the diameter of semicircle of impedance spectra are obtained as 2.3 kΩ for blank ITO electrode (curve a), 2.77 kΩ (curve b) for the Nano-CeO_2/ITO electrode and 2.92 kΩ for Urs-GLDH/Nano–CeO_2/ITO bioelectrode, respectively. The increase of R_{ct} (2.92 kΩ) value and the shift of semicircle to higher frequency after deposition of the Nano–CeO_2 layer on the ITO electrode is attributed to the low electrical conductivity of the Nano–CeO_2 particles that provides slow electron transfer between Nano-CeO_2 particles and electrode. This result suggests that solution resistance (R_s) is slightly increased, but interfacial impedance (double layer capacitance, C_{dl} and the charge transfer resistance, R_{ct}) is greatly increased after the deposition of Nano-CeO_2 film on the ITO electrode surface. It is observed that after the immobilization of mixed Urs-GLDH enzyme on the matrix the R_{ct} value is increases dramatically due to insulating characteristics of the mixed Urs-GLDH enzyme molecules. It reveals that Warburg impedance (D), which represents the mass transfer, is enhanced after the immobilization of these enzymes (Urs-GLDH) on the Nano–CeO_2/ITO electrode surface, implying that enzyme molecules are indeed absorbed on Nano–CeO_2/ITO surfaces and induce impedance effects on the interface of the electrode surface. The double-layer capacitance (C_{dl}) is decreased, indicating that a higher enzyme (Urs-GLDH) concentration enhances the adsorption of enzyme on the Nano–CeO_2/ITO modified electrode.

Figure 3. SEM micrographs of (a) sol-gel derived Nano-CeO_2 electrode and (b) Urs-GLDH/Nano-CeO_2/ITO bioelectrode

3.5. Rlectrochemical impedance spectroscopy

The changes of electrode behavior after surface modification with enzymes (Urs and GLDH) were studied by cyclic voltammetry (CV) in the presence of ferricyanide mediator. When the electrode surface is modified by the biocatalytic material, the change in the electron transfer kinetics of $[Fe(CN)_6]^{3-/4-}$ gives indication of the enzyme attachment. Figure 5 shows the cyclic voltammograms of blank ITO (a), Nano-CeO_2/ITO (b) and Urs-GLDH/Nano-CeO_2/ ITO bioelectrode (c) at a scan rate of 50 mVs^{-1}. As shown in Figure 5 well-defined redox couple of $[Fe(CN)_6]^{3-/4-}$ is observed on the bare ITO electrode (curve a). When the Nano–CeO_2/ITO electrode layer is deposited on the ITO electrode, the peak current gradually decreases (curve b). The decrease in the peak potential after deposition of Nano–CeO_2 may be attributed to the hindrance of the electrons flow as a result of reduction in electrical conductivity of the nanostructured Nano–CeO_2/ITO electrode. Further, redox current of Urs-GLDH/Nano–CeO_2/ITO bioelectrode is found to decrease due to insulating characteristics of Urs-GLDH indicating slow down of redox process during the biochemical reaction. The reason is that an insulating layer of non-conducting enzyme had been assembled on electrode surface, which act as a electron transfer barrier.

Figure 6 demonstrates typical CV of Urs-GLDH/Nano-CeO_2/ITO bioelectrode with scan rate varying from 10 to 100 mVs^{-1}. With increase in the scan rate, there is increase in both the cathodic and anodic peak current accompanied with small shift and increased peak-to-peak separation. The cathodic and anodic peak currents increase linearly with scan rate varying from 10 to 100mVs^{-1} indicating a surface controlled electrode process. This reveals that the electron transfer between enzyme and electrode could be easily performed and it was a surface confined electrochemical process.

Journal of Physics: Conference Series **358** (2012) 012006 doi:10.1088/1742-6596/358/1/012006

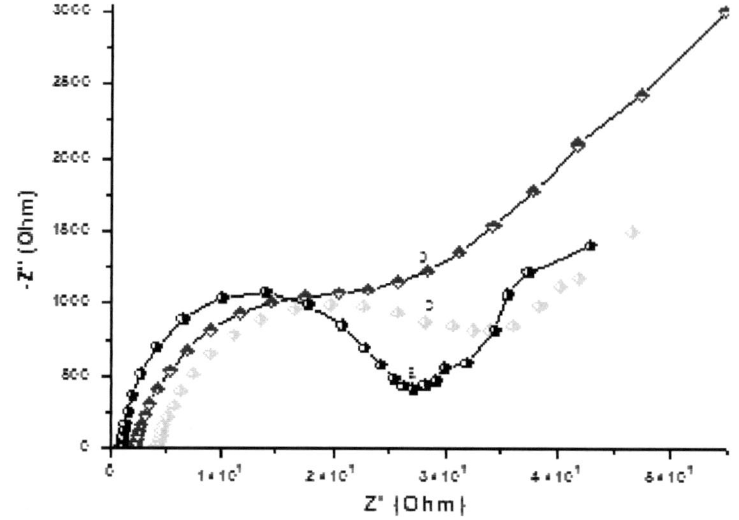

Figure 4. Electrochemical Impedance spectroscopy of (a) bare ITO electrode (b) sol-gel derived Nano-CeO_2/ITO electrode and(c) Urs-GLDH/Nano- CeO_2/ITO bioelectrode in PBS solution (50 mM, pH 7.0, 0.9% NaCl) containing 5 mM $[Fe(CN)_6]^{3-/4-}$

3.6. Rlectrochemical impedance spectroscopy

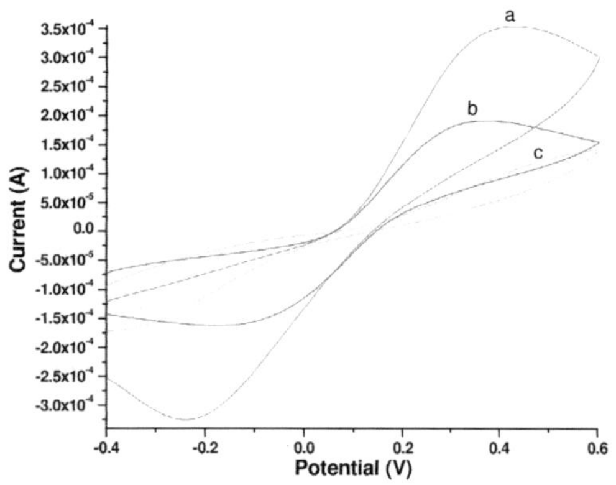

Figure 5. Cyclic voltammogram of (a) bare ITO electrode (b) sol-gel derived Nano-CeO_2/ITO electrode and(c) Urs-GLDH/Nano- CeO_2/ITO bioelectrode in PBS solution (50 mM, pH 7.0, 0.9% NaCl) containing 5 mM $[Fe(CN)_6]^{3-/4-}$

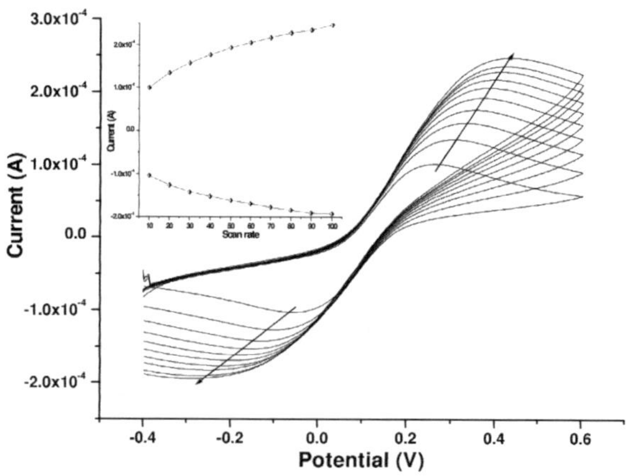

Figure 6. CV of Urs-GLDH/Nano-CeO_2/ITO bioelectrode as a function of scan rate (10-100 mV/s^{-1}) in ascending order in PBS solution containing 5 mM $[Fe(CN)_6]^{3-/4}$

Figure 7 (inset) illustrate the CV response studies of the Urs-GLDH/Nano-CeO_2/ITO bioelectrode in PBS solution {pH 7, 0.9%NaCl, containing 5mM $[Fe(CN)_6]^{3-/4-}$} in the presence of 30 μL of nicotinamide adenine dinucleotide (NADH, 3.7 mg/dL) and 70 μL of α-Keto glutamate (α-KG, 47.5

mg/dL) and a successive addition of urea at an applied 50 mV/s scan rate. The peak current rises sharply with increased concentration of urea, with the maximum response approach at the concentration of 80 mg/dL. This increase in electrochemical current response can be attributed to pH sensitive behavior of the Ur-GLDH/Nano-CeO_2/ITO bioelectrode. The sensor achieves 95% of the steady-state current in less than 5s indicating fast electron exchange between the Urs-GLDH and Nano–CeO_2/ITO electrode. Figure 7 shows the amperometric current of the fabricated biosensor increases linearly within the concentration range 10-80 mg/dL, with increasing urea concentration the correlation coefficient *(R)* as 0.991 indicating good electro-catalytic behavior of Urs-GLDH/Nano–CeO_2/ITO bioelectrode. The sensitivity of the Urs-GLDH/Nano–CeO_2/ITO bioelectrode calculated from the slope of curve was found to be 0.64 μA/mg/dL and low detection limit 0.1 mg/dL.

The value of K_m, which is a reflection of enzymatic affinity, was found to be 6.09 mg/dL. The lower K_m value with respect to glutaraldehyde-cross-linked urease-albumin gel (6.5 mM) and TMOS-based sol-gel (4.95 mM) indicates easier diffusion of substrate and product molecules into and out of the Nano–CeO_2 film and the feasible configuration of the enzyme in the nanomaterial matrix. The high affinity of Urs-GLDH to urea can be attributed to Nano–CeO_2 nanoparticles because of its biocompatibility, large surface area, and high electron communication capability. The value of ionization electron potential (IEP) of CeO_2 and urease are known to be 9.0 and 5.3, respectively. Therefore, the Nano–CeO_2 matrix is positively charged in contrast to the enzymes that are negatively charged. Thus, immobilization of Urs-GLDH on the Nano–CeO_2 nanoparticles matrix is highly favored via electrostatic interaction.

3.7. Effect of pH onto Urs-GLDH/nano-CeO_2/ITO biosensor

The influence of pH of buffer on the response of the biosensor studied over the range 6.0–8.0 indicates that the current response increases with increase of pH of the substrate solution up to 7.0, after which it decreases. In view of the prime pH of the enzyme, the neutral pH was selected throughout this work (Figure 8).

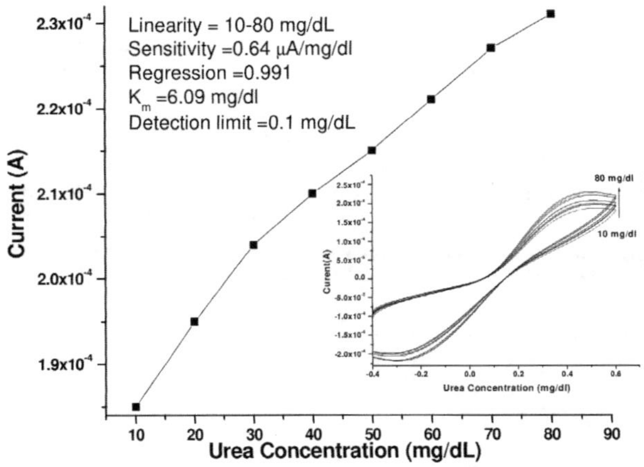

Figure 7. The calibration plot between current response and different concentration of urea in the range (10-80 mg/dl)in PBS solution containing 5 mM $[Fe(CN)_6]^{3-/4-}$; inset shows the CV response of Urs-GLDH/Nano-CeO_2/ITO bioelectrode as a function of urea concentration

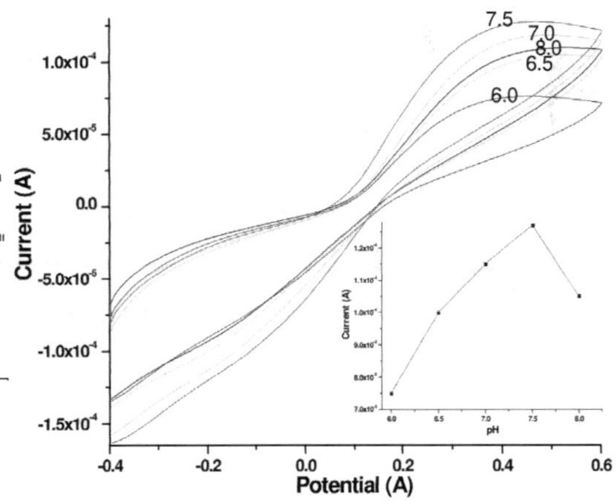

Figure 8. Effect of pH on Urs-GLDH/ Nano-CeO_2/ITO bioelectrode using CV

4. Conclusions

Nanostructured CeO_2 film was prepared by sol gel technique and used to co-immobilization of Urs and GLDH enzymes via physical adsorption for urea detection. This biosensor exhibits excellent performance characteristics, such as sensitivity (0.64μA/mg/dL) and reproducibility, wide linear range (10–80 mg/dL), low detection limit (0.10 mM), and long-term stability of about 6 months. A relatively low value of the Michaelis–Menten constant obtained as 6.09 mg/dL indicates enhanced enzyme affinity of Urs to urea. The wide range of detection and high sensitivity may be assigned to amplification of the magnitude of current due to the alignment of Nano–CeO_2 nanoparticles to the matrix. Efforts should be made to use this electrode for the detection of urea in blood serum.

References

[1] Wang J X, Sun X W, Wei A, Lei Y, Cai X P, Li C M and Dong Z L 2006 *Appl. Phys. Lett*. **88** 233106
[2] Pandey P, Datta M and Malhotra B D 2008 *Anal. Lett.* **41** 157–207
[3] Zhou H, Gan X, Wang J, Zhu X and Li G 2005 *Anal. Chem.* **77** 6102
[4] Kumar S A and Chen S M 2008 *Anal. Lett.* **41** 141
[5] Singh S P, Arya S K, Pandey M K, Malhotra B D, Saha S, Sreenivas K and Gupta V 2006 *Appl. Phys. Lett.* **91** 063901
[6] Wei A, Sun X W, Wang J X, Lei Y, Cai X P, Li C M, Dong Z L and Huang W 2006 *Appl. Phys. Lett.* **99** 123902
[7] Wang S G, Zhang Q, Wang R and Yoon S F 2003 *Biochem. Biophys. Res. Commun.* **311** 572–576
[8] Galatanu A N, Chrorrakis I S, Anghel D F and Khan A 2000 *Langmuir* **16** 4922
[9] Shao Y, Wang J, Wu H, Liu J, Aksay I A and Li Y 2010 *Electroanalysis* **22** 1027 – 103
[10] Wanekaya A K, Chen W, Myung N Vand Mulchandani A 2006, Electroanalysis **18** 533 – 55
[11] Khan R, Kaushik A, Solanki P R, Ansari A A, Pandey M K and Malhotra B D 2008 *Analytica Chimica Acta* **616** 207–213
[12] Ansari A A, Kaushik A, Solanki P R, and Malhotra B D 2008 *Electrochem. Commun.* **10** 1246
[13] Solanki P R, Kaushik A, Ansari A A, Sumana G, and Malhotra B D 2008 *Appl. Phys. Lett.* **93** 163903
[14] Suzuki T, Kosacki I and Anderson H U 2001 *J. Am. Ceram. Soc.* **84** 2007
[15] Ghodsi F E, Tepehan F Z and Tepehan G G 1999 *Electrochim. Acta* **44** 3127
[16] Jia N Q, Zhou Q, Liu L, Yan M M, and Jiang Z Y 2005 *J. Electroanal. Chem.* **580** 213
[17] Wang J 1999 *Anal. Chim. Acta* **399** 21-27
[18] Lev O, Wu Z, Bharathi S, Glezer V, Modestov A, Gun J, Rabinovich L and Sampath S 1997 *Chem. Mater.* **9** 2354-237
[19] **Ali A,** Ansari A A, Kaushik A, Solanki P R, Barik A, Pandey M K and Malhotra B D 2009 *Materials Letters* **63** 2473-2475
[20] Cho W-J and Huang H-J 1998 *Anal. Chem.* **70** 3946-395
[21] Vostiar I, Tkac J, Sturdik E and Gemeiner P 2002 *Bioelectrochemistry* **56** 113 – 115
[22] Saha S, Arya S K, Singh S P, Sreenivas K, Malhotra B D and Gupta V 2009 *Biosen. & Bioelectron.* **24** 2040-2045
[23] Feng K J, Yang Y H, Wang Z J, Jiang J H, Shen G L and Yu R Q 2006 *Talanta* **70** 561

Acknowledgements

We thank the Director, NPL, New Delhi, India for the facilities. Financial support received under the DBT project GAP-070832, DST project GAP- 070932 and in-house project OLP-070632D is gratefully acknowledged.

Journal of Physics: Conference Series **358** (2012) 012007 doi:10.1088/1742-6596/358/1/012007

Opportunities in nano-structured metal oxides based biosensors

B D Malhotra*[1, 2] Maumita Das[1] and Pratima R Solanki[1]
[1]Department of Science & Technology Centre on Biomolecular Electronics, Biomedical Instrumentation Section, National Physical Laboratory, Dr. K. S. Krishnan Marg, New Delhi 110012, India
[2]Department of Biotechnology, Delhi Technological University, Shahbad Daulatpur, Main Bawana Road, Delhi 110042, India

E-mail: bansi.malhotra@gmail.com or bansi.malhotra@dce.ac.in

Abstract. Nanomaterials are presently at the critical stage of the next technological revolution in solid-state electronics and are emerging as new structural materials, to serve as systems for controlled drug delivery, biomolecular electronics and are considered to have considerable impact in practically all domains of science. Among the various types of nanomaterials that have been developed, nanostructured metal oxides (NSMOs) have recently aroused much interest as immobilizing matrices for biosensors development. The unique properties of NSMOs offer excellent prospects for interfacing biological recognition events with electronic signal transduction and for designing a new generation of bioelectronics devices that may exhibit novel functions. Among the NSMOs, biocompatible zirconia (ZrO_2) and its composite especially with chitosan and carbon nanotubes are technologically important exhibits high bioactivity for biomolecules with rapid rand enhanced electrochemical signal. Efforts are being continuously made to explore the prospects and future challenges of NSMOs for the development of biosensing devices.

1. Introduction

Nano-structured materials can be grouped as (i) inorganic, (ii) organic and (iii) inorganic-organic nanocomposites. Inorganic nanostructures especially nano-structured metal oxides (NSMOs) e.g. iron oxide (Fe_2O_3), zirconium oxide (ZrO_2), zinc oxide (ZnO), nickel oxide (NiO), cerium oxide (CeO_2) silicon dioxide (SiO_2) etc, have opened new opportunities for a plethora of applications due to their unique properties (Figure 1) [7-11]. Electron transport properties of NSMOs are very important for electrical and electronic applications as well as for understanding the unique one-dimensional carrier transport mechanism. Besides this, metal oxide nanoparticles have high thermal stability, chemical inertness, non-toxicity, large surface-to-volume ratio, high surface reaction activity, biocompatibility and tunable electron transport properties. However, these NSMOs are difficult to functionalize that can perhaps be addressed by combining them with organic nanostructures which are otherwise not stable alone, as they interact with the moieties present in the environment. There is, thus a synergic effect between the inorganic and organic nanostructure and these inorganic-organic nanocomposite

* To whom any correspondence should be addressed.

offer excellent prospects for interfacing biological recognition events with electronic signal transduction and for designing new biosensor devices [12-15].

According to IUPAC, a biosensor is a self-contained integral device that is capable of providing specific quantitative or semi-quantitative analytical information using a biological element. Biosensor has three major components: (i) a bio-recognition element such as DNA, enzyme, antibody etc. for recognition of an analyte also called bioreceptor (ii) an immobilization surface such as a nanomaterial/conducting polymer/sol–gel film/self-assembled monolayer, etc., used for the immobilization of a biomolecule and (iii) a transducer for conversion of a biochemical reaction product into a recognizable signal.

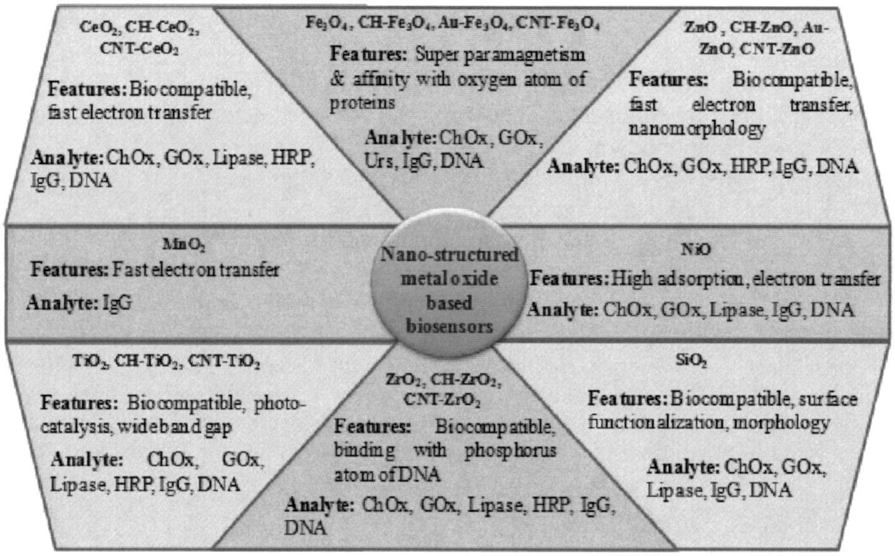

Figure 1. Properties of some NSMOs used for the fabrication of biosensors.

2. Importance of NSMOs for biosensing

Biosensor is based on the interaction of a biomolecule in conjunction with an electrode where the redox reaction could be monitored electrochemically by measurement of the loss or formation of a substrate or product, by the use of a small mediator species that shuttles between the biomolecule and the electrode, or less commonly by direct electron transfer (ET) between biomolecule redox site and the electrode. Direct ET can be difficult to achieve, since the redox site of a biomolecule is often hidden deep inside the biomolecule. Recent advances in achieving direct ET have been made by the modification of biomolecules or electrode surface through the use of novel NSMO as a mediator and the design of functional biointerface (Figure 2), which has extended throughout the field of bioelectrochemistry [16-19].

The NSMOs have recently become important as immobilization matrices for the fabrication of biosensors, as they provide excellent optical and electrical properties due to electron and phonon confinement, high surface-to-volume ratio, high surface reactivity and catalytic activity along with strong adsorption ability resulting in enhanced sensing characteristics. The adsorption of biomolecules directly onto the bare surface a of bulk material may frequently result in their denaturation and loss of bioactivity. However, the adsorption of such biomolecules onto a given NSMO surface may perhaps retain the bioactivity because of the non-toxicity NSMO and their biocompatible composites with carbon nanotubes (CNT), chitosan (CHIT) etc using electrostatic and other interactive forces.

The interface formed between an NSMO and a biomolecule is of utmost importance as the surface area, surface energy, roughness, porosity etc. affect the performance of a biosensor. So, the judicious selection of NSMO is the key success for the development of an effective biosensor [20].

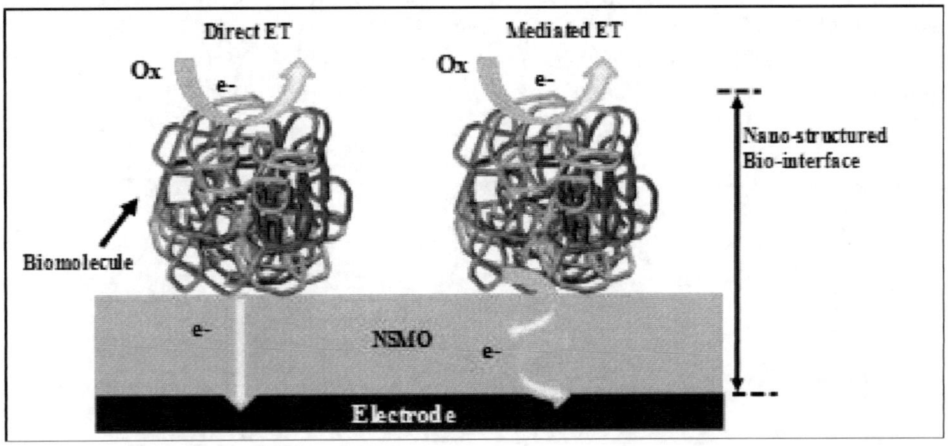

Figure 2. Scheme of ET process of nanostructured bio-interfaces on the electrode surface.

An effective NSMO biointerface helps a biomolecule, to retain its biological activity with high stability by establishing a biocompatible microenvironment, and also faster rate of electron transfer. There is thus, a considerable opportunity for the development of biosensors with improved sensitivity and detection limits, as well as lower cost and extended shelf life, through the use of NSMOs.

3. Electrochemical DNA biosensor using nano-structured biocompatible ZrO_2 based matrices

Fascinating electrochemistry and unique hybridization behaviour of desired nucleic acid and the excellent prospects for interfacial biological recognition event with electronic signal transduction using NSMOs, has stimulated much interest for designing new generation bioelectronic devices that may exhibit novel applications especially in the field of DNA biosensors. The high sensitivity, specificity for target analyte, faster response, ability to operate in turbid or colored solutions, easy handling, low cost, compatibility with micro-fabrication technology and portability has aroused increased interest for the fabrication of DNA biosensors [21-23].

For the fabrication of an electrochemical DNA biosensor, immobilization of DNA onto a nanocrystalline transparent metal oxide based matrices like ZrO_2 have drawn considerable attention due to their unique physical, chemical, and optical properties that make them promising matrices for sensing applications. Moreover, the nanostructures provide increased surface area for DNA immobilization that may lead to the improved limit of detection. Besides this, it is found to be thermally stable, chemically inert, non-toxic, and has affinity for groups containing oxygen that facilitates covalent immobilization of biomolecules without using any cross-linker that may limit sensitivity of a biosensor. The biocompatibility of ZrO_2 is anticipated to offer conducive environment to the fabricated biosensor which may perhaps result in enhanced stability and reusability.

Solanki et.al have utilized sol–gel-derived nano-structured ZrO_2 film for immobilization of 17-base ssDNA identified from the 16srRNA coding region of *Escherichia coli.* This bioelectrode exhibits high selectivity and sensitivity for hybridization, with linearity in the range of 10^{-6} to 10^{6} pM of complementary DNA [24]. *Zuo et al.* have fabricated a DNA biosensor by immobilizing ssDNA with a terminal 5'-phosphate group to a ZrO_2 surface utilizing the strong affinity between the oxygen atoms of the phosphoric group and zirconium. This biosensor has been found to have high sensitivity and selectivity for hybridization detection (≤20 nM complementary DNA) [25]. Yang et al. have fabricated a DNA biosensor using a composite comprising of Nano ZrO_2-CNT-CHIT and observed that the bioelectrode can effectively detect the complementary DNA 1.49×10^{-10} – 9.32×10^{-8} using differential pulse voltammetry [26].

Inspired by the characteristics of nano ZrO_2, *Das et.al* have fabricated electrochemical DNA biosensor for *M. tuberculosis* detection using nano-structured ZrO_2, from an aqueous solution of zirconium oxychloride ($ZrOCl_2$) onto gold coated glass plate (Au) (Figure 3). They have observed that ssDNA-NanoZrO_2/Au electrode can detect target analyte from 640-0.065 ng.µL^{-1}, with lower detection limit of 0.065 ng.µL^{-1}(20nM) and 1 ng.µL^{-1} for genomic DNA indicating that it can be used for rapid and early detection of *M. tuberculosis* [27].

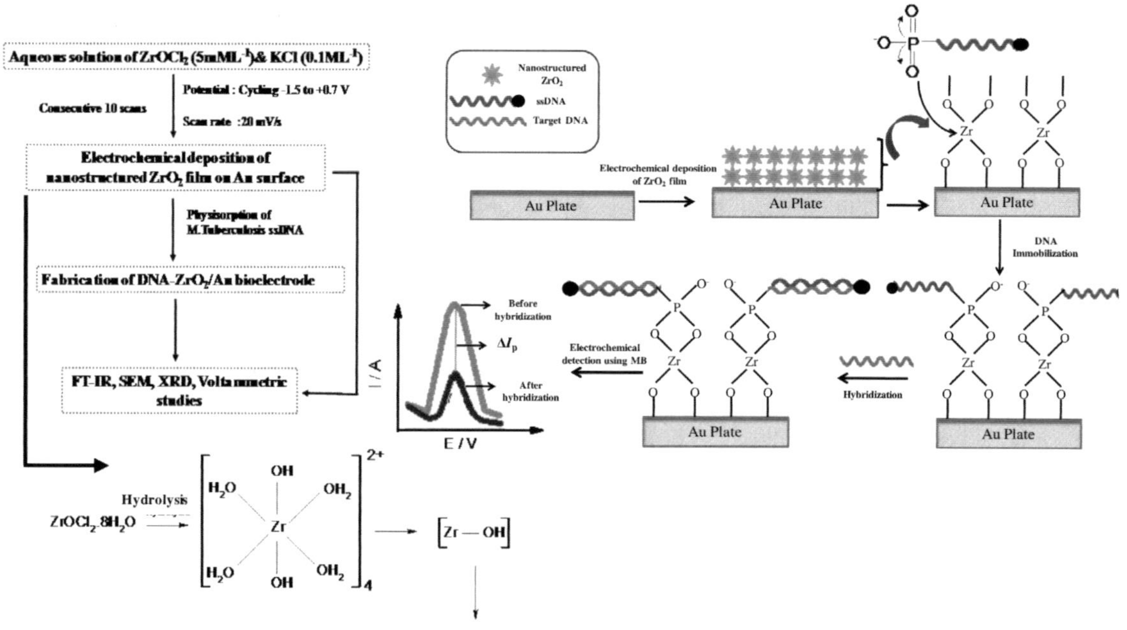

Figure 3. Proposed schematic for the fabrication of NanoZrO_2/Au based DNA biosensor.

In spite of being an interesting matrix for biomolecule immobilization, nano-structured ZrO_2 based electrode suffers from the problems of cracking and aggregation leading to limited application of ZrO_2 nanoparticles to biosensing. This problem can perhaps be addressed by modifying ZrO_2 nanoparticles with CHIT. CHIT has been found to be an interesting biopolymer for immobilization of biomolecules because of its excellent film-forming ability, high permeability, biocompatibility, low cost etc. Moreover, chemical modification of amino groups of CHIT provides hydrophilic environment for the biomolecules. *Feng et al.* have reported nanoporous CeO_2/CHIT composite film as the immobilization matrix for fabrication of electrochemical DNA biosensor for colo-rectal cancer detection [28]. This biocomposite matrix can be used for increased loading of probe DNA to obtain enhanced biosensor response. *Kaushik et al.* have developed a nucleic acid sensor based on Fe_2O_3/CHIT hybrid matrix for pyrethroid detection [29].

We have proposed chitosan-zirconium-oxide (CHIT-NanoZrO_2) nanocomposite, deposited via electrophoretic deposition of onto indium-tin-oxide(ITO) coated glass plate from a colloidal suspension of CHIT-NanoZrO_2 in acetonitrile, to reveal its application in DNA biosensor for *M. tuberculosis* detection. Efforts have been directed to propose a mechanism for the electrophoretic deposition of CHIT-NanoZrO_2 colloidal particles onto given ITO plate in presence of DC electric field. A key characteristic of CHIT, used for the preparation of CHIT-NanoZrO_2 composite film fabrication, is its unique response to applied electrical stimuli. When applied voltage is sufficient for protons to be reduced at the cathode surface, a localized pH gradient is generated, resulting in cathodic electrodeposition of thin CHIT-NanoZrO_2 film (180 nm) with ZrO_2 molecules entrapped within CHIT chains due to strong electrostatic interactions between these oppositely charged moieties [30].

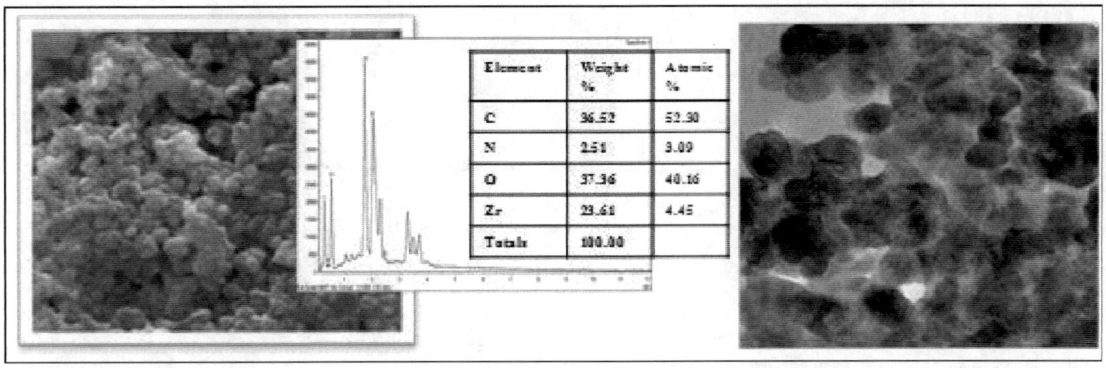

Element	Weight %	Atomic %
C	36.52	52.30
N	2.51	3.09
O	37.36	40.16
Zr	23.61	4.45
Totals	100.00	

Figure 4. SEM and TEM analysis of CHIT-NanoZrO_2 composite

The morphological studies carried out using SEM and TEM techniques clearly reveal incorporation of the hexagonal grains of ZrO_2 (30-50 nm) in the amorphous network of CHIT (Figure 4). Interestingly, the nano grained ZrO_2 have been found to be inter-linked and well-dispersed into the CHIT matrix, corroborating the homogeneous nanobiocomposite formation. This has been attributed to electrostatic interactions between cationic CHIT and surface charged ZrO_2 nanoparticles. The DNA/CHIT-NanoZrO_2/ITO bioelectrode, fabricated by covalent immobilization of biotinylated probe DNA specific to *M. tuberculosis* onto this nanobiocomposite matrix, has been found to be selective and can detect complementary target up to 0.78 nM with sensitivity of 6.38×10^{-6} A μM^{-1}.

Contemporary interest in metal and metal oxide/CNT composites stems from their utilization as an emitter in field emission displays, a thermal interface material, a super capacitor, interconnects in large scale integrated circuits etc [31-32]. Keeping in view, the advantages of CNT-metal oxide composites in various emerging fields, there is an augmented interest in exploring the synergy of electroactive CNT and biocompatible metal oxides in the potential field of biomolecular electronics. In this context, bio-functional devices based on CNT/ZrO_2 nanocomposites have recently gained much attention [33-35]. In the fabrication of a biosensing device composed of CNT/ZrO_2, the thickness and the morphology of the ZrO_2 coating is considered to play a critical role. The uniform, dense pin-hole free thin coating provides large surface area that can be useful for biosensor applications. In this context, electrophoretic technique has achieved much interest for deposition of the uniform, dense and electrochemically reversible films from colloidal suspension [36]. In addition, this technique provides a unique strategy to prepare nano-patterned composite film with nanoporosity and densely packed structure with controlled thickness and morphology.

Electrophoretic deposition of ZrO_2 grafted CNT, has been obtained by *Das et al.* via isothermal hydrolysis of zirconium oxychloride ($ZrOCl_4$) in presence of CNT, onto ITO coated glass plate (Figure 5). The structural characterization of this nanocomposite (Figure 6) divulge that there is transition of crystal structure of ZrO_2 from its pristine monoclinic to deviated monoclinic arrangement with increased unit cell volume of 163 $Å^3$. The high resolution TEM and SEM studies reveal that the ZrO_2 is present in and around the nanotubes resulting in deviated monoclinic crystalline arrangement having crystallite size of 28.63 nm. The change in the crystalline structure provides increased surface area to immobilize the biomolecules required for the fabrication of an effective biosensor.

Based on this idea, an impedimetric genosensor (ssDNA-NanoZrO_2-CNT/ITO) is fabricated by *in-situ* entrapment of 21-mer DNA probe specific to *Mycobacterium tuberculosis* [37].

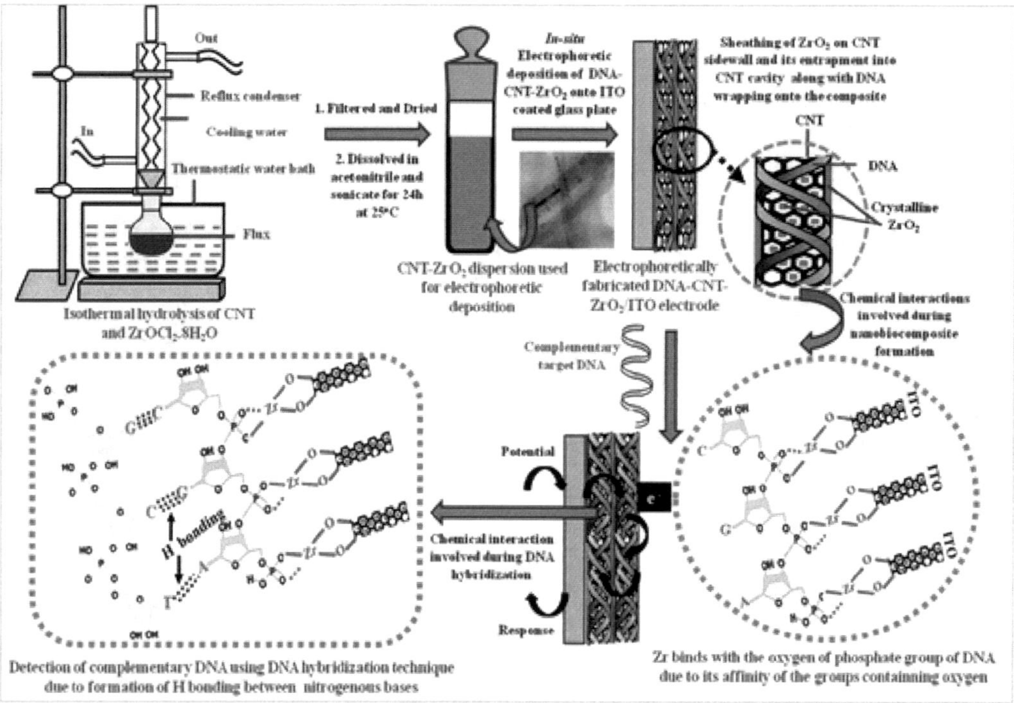

Figure 5. Electrophoretic fabrication of CNT/ZrO_2 nanocomposite and the interaction between DNA and the nanocomposite.

(A)

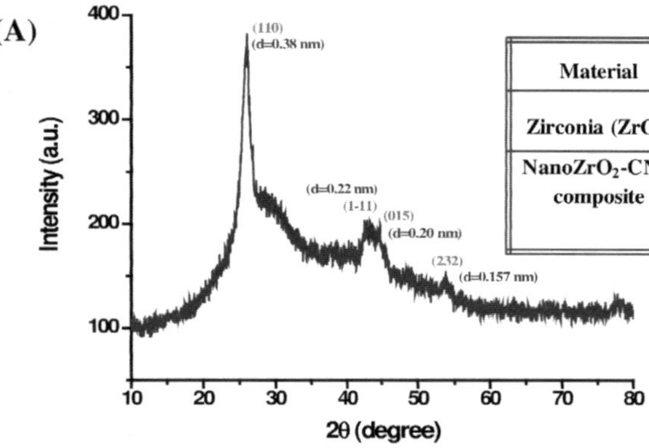

Material	Lattice parameter	Unit cell volume
Zirconia (ZrO_2)	a = 5.1495; b = 5.2075; c = 5.3164 α = β = 99.23 ; γ	140.7188 Å^3
NanoZrO_2-CNT composite	a = 5.9128; b = 5.8082; c = 4.8088 α = 90.000 ; β = 90.943 ; γ = 90.000	163.139 Å^3

Figure 6. (A) XRD; (B) TEM ; (C) SEM and (D) EDX pattern of NanoZrO_2-CNT composite.

The ssDNA-NanoZrO_2-CNT/ITO electrode is capable of detecting DNA hybridization process by showing the variation in charge transfer resistance (R_{CT}) obtained from the diameter of the semicircle of the Nyquist plot. The results of DNA hybridization with the complementary target DNA strands reveal wider detection range of target DNA concentration (1×10^{-2} mM to 1×10^{-8} mM) and exhibit detection limit as low as 1×10^{-8} mM or 0.01 nM with storage stability of 24 weeks.

Table 1 exhibits the various biosensing characteristics obtained for the fabrication of DNA biosensors for *M. tuberculosis* detection using ZrO_2 and its composite by *Das et al.*

Table 1. Various biosensing parameter obtained using ZrO_2 and its composite with CHIT and CNT

Biosensing characteristics / Matrix	Film fabrication method	Detection limit	Reusability	Stability
NanoZrO_2	**Electrochemical**	**20 nM**	**10-12 times**	**16 weeks**
NanoZrO_2-CHIT	**Electrophoretic**	**0.78 nM**	**10-12 times**	**18 weeks**
NanoZrO_2-CNT	**Electrophoretic**	**0.01 nM**	**~ 15 times**	**24 weeks**

4. Conclusions

With the rapid emergence of nanotechnology, hybrid bio/nanostructures with well-defined morphologies have been popularly employed to modern materials science. Capitalizing the properties of NSMOs, novel composite nano-scale materials have opened up a new horizon in materials science with a synergistic effect of the individual components. The highly sensitive detection protocol for the fabrication of biosensor using NSMOs especially with ZrO_2, suggest that future interdisciplinary research is likely to lead to a new generation of electrochemical biosensors for the detection of infectious agent in an early stage.

References

[1] Gooding J J 2006 *Small* **2** 313
[2] Caruso F 2001 *Adv. Mater.* **13** 11
[3] Rosenholm J M, Meinander A, Peuhu E, Niemi R, Eriksson J E, Sahlgren C and Linden M 2009 *ACS Nano* **3** 197
[4] Ghadiali J E and Stevens M M 2008 *Adv. Mater.* **99** 1
[5] Forzani E S, Zhang H Q, Nagahara L A, Amlani I, Tsui R and Tao N J 2004 *Nano Lett.* **4** 1785
[6] Wang J X, Sun X W, Wei A, Lei Y, Cai X P, Li C M and Dong Z L 2006 *Appl. Phys. Lett.* **88**, 233106.
[7] Liu Z, Liu Y, Yang H, Yang Y, Shen G and Yu R 2005 *Electroanalysis* **17** 1065
[8] Wei H and Wang E 2008 *Anal. Chem.* **80** 2250
[9] Kim H J, Yoon S H, Choi H N, Lyu Y K and Lee W Y 2006 *Bull. Korean Chem. Soc.* **27** 65
[10] Xiao X, Luan Q, Yao X and Zhou K 2009 *Biosens. Bioelectron.* **24** 2447
[11] Feng K J, Yang Y H, Wang Z J, Jiang J H, Shen G L and Yu R Q 2006 *Talanta* **70** 561
[12] Hook F, Kasemo B, Grunze M and Zauscher S 2008 *ACS Nano* **2** 2428
[13] Hu c and Hu S 2009 *J. Sens.* **1** 1
[14] Nel A E Mädler L, Velegol D, Xia T, Hoek Eric M V, Somasundaran P, Klaessig F, Castranova V and Thompson M 2009 *Nat. Mater.* **8** 543
[15] Rahman M M, Ahammad A J S, Jin J H, Ahn S J and Lee J J 2010 *Sensors* **10** 4855
[16] Katz E, Willner I and Wang J 2004 *Electroanalysis* **16** 19
[17] Hartmann M 2005 *Chem. Mater.* **17** 4577
[18] Herna´ndez-Santos D, Gonzalez-Garcıa M B and Garcıa A C 2002 *Electroanalysis* **14** 1225

[19] Wang J 2005 *Electroanalysis* **17** 7
[20] Solanki P R, Kaushik A, Agrawal V V and Malhotra B D 2011 *NPG Asia Mater.* **3** 17
[21] Vaníková M, Lehotay J, Imárik J and Labuda J 2005 *Bioelectrochemistry* **66** 125
[22] Palecek E andFojta M 2001 *Anal. Chem.* **73** 74A
[23] Burda C, Chen X, Narayanan R and El-Sayed M A 2005 *Che.Rev.* **105** 1025
[24] Solanki P R, Kaushik A, Chavhan P M, Maheshwari S N and Malhotra B D 2009 *Electrochem. Commun.* **11** 2272
[25] Zuo S-H, Zhang L F, Yuan H H, Lan M B, Lawrance G A and Wei G 2009 *Bioelectrochemistry* **74** 223
[26] Yang Y, Wang Z, Yang M, Li J, Zheng F, Shen G and Yu R 2007 *Anal. Chim. Acta.* **584** 268
[27] Das M, Sumana G, Nagarajan R and Malhotra B D 2010 *Appl.Phs. Lett.* **96** 133703
[28] Feng K J, Yang Y H, Wang Z J, Jiang J H, Shen G L and Yu R Q 2006 *Talanta* **70** 561
[29] Kaushik A, Solanki P R, Ansari A A, Malhotra B D and Ahmad S 2009 *Biochem. Eng. J.* **46** 132
[30] Das M, Dhand C, Sumana G, Srivastava A K, Nagarajan R, Nain L, Iwamoto M, Manaka T and Malhotra B D 2011 *Biomacromolecules* **12** 540
[31] Weizmann Y, Chenoweth D M and Swager T M 2011 *J. Am. Chem. Soc.***133** 3238
[32] Nugent J M, Santhanam K S V, Rubio A and Ajayan P M 2001*Nano Lett.* **1** 87
[33] Kim I H, Kim J H, Cho B W, Lee Y H and Kim K B J 2006 *Electrochem. Soc.***153** A989
[34] Brasse G, Restoin C, Soule D and Blondy J M 2011 *J. Phys.Chem.C.* **115** 248
[35] Meighan M M, Vasquez J, Dziubcynski L, Hews S and Hayes M A 2011 *Anal. Chem.* **83** 368
[36] Tada K and Onoda M *Adv Funct Mater.*, 2002 **2** 420
[37] Das M, Dhand C, Sumana G, Srivastava A K, Vijayan N, Nagarajan and Malhotra B D 2011 *Appl.Phy. Lett.* **99** 143702

Journal of Physics: Conference Series **358** (2012) 012008 doi:10.1088/1742-6596/358/1/012008

Fabrication of nanocrystalline CdS electrode via chemical bath deposition technique for application to cholesterol sensor

Hemant Dhyani[1], Saurabh Srivastava[2], Md. Azahar Ali[2], B.D. Malhotra[2, 3] and Prasenjit Sen[1]
[1] School of Physical Sciences, Jawaharlal Nehru University, New Delhi-110067, India
[2]Department of Science and Technology Centre on Biomolecular Electronics, National Physical Laboratory, Dr. K. S. Krishnan Marg, New Delhi-110012, India
[3]Deapartment of Biotechnology, Delhi Technological University, Shahbad Daulatpur, Main Bawana Road, Delhi, 110042, India

E-mail: "Prasenjit Sen" psen.jnu@gmail.com

Abstract. A nanocystalline CdS electrode has been fabricated by chemical bath deposition (CBD) technique onto hydrolyzed indium tin oxide (ITO) coated glass substrate at 78^0C for the immobilization of cholesterol oxidase (ChOx). The prepared Nano-CdS based electrode has been characterized using UV-visible, X-ray diffraction (XRD), Fourier transform-infrared (FTIR) and scanning electron microscopy (SEM). The ChOx/Nano-CdS/ITO bioelectrode shows the detection range of cholesterol from 50 to 400 mg/dl with improved sensitivity of 1.35 μA/mgdl^{-1}/cm^2, low detection limit (6.1 mg/dl) and low K_m (0.45mM) value indicating strong enzyme (cholesterol oxidase)-matrix (CdS) affinity.

1. Introduction

In recent years semiconductor quantum dots (Qdots) have attracted much interest in the different fields of science due to their unique optical, electronic and electrochemical properties. These properties especially large absorption spectra but narrow emission bands, excellent photostability, high quantum yield and size dependent photoluminescence make them desirable fluorescence probes for the sensing of biological samples [1-4]. Apart from optical sensing, other platforms comprising of quantum dots have recently been explored for the sensing of biological analytes using the electrochemical technique .In this context, it has been reported that the semiconductor Qdots can be used to obtain enhanced charge transfer in enzymatic electrochemical biosensor [5-9]. The Qdots deposited onto desired substrate, have been used for H_2O_2 estimation [10]. Vastarella *et. al.* have fabricated enzyme-core/shell nanoparticle hybrid material and enzyme/semiconductor nanocluster for the amperometric biosensing [11]. Among the various semiconductors Qdots, the CdS nanocrystals have attracted considerable attention due to their good chemical stability, photo-electrochemistry, intrinsic properties of a narrow band gap and excellent electro-catalysts for the sulfide/polysulfide redox couple [12-14]. The good electron transport behavior of CdS quantum dots due to higher charge detaching efficiency [15] and high enzyme affinity to the hexagonal nanocystalline surface area has recently been reported for electrochemical sensing [16,17]. There are several methods to deposit the semiconductor CdS onto a substrate such as electrodeposition, vacuum evaporation, photochemical

[1] To whom any correspondence should be addressed.

deposition, spray pyrolysis, sputtering , Chemical bath deposition (CBD) etc. The CBD technique is an easy process to achieve high quality crystalline films by adjusting the temperature, reagent concentration and pH for large area industrial applications with low cost.

We report results of the studies relating to preparation of nanocystalline CdS film onto indium tin oxide coated glass substrate by chemical bath deposition (CBD) technique for the immobilization of cholesterol oxidase used for cholesterol estimation.

2. Experimental

2.1. Fbrication of the nano-CdS/ITO electrode

The CdS quantum dots have been prepared and deposited simultaneously onto ITO substrate by chemical bath deposition (CBD) technique reported in literature [18] with some modifications. For the synthesis of CdS, 36.60mg cadmium chloride, 152.24mg thiourea and 1.2gm ammonium nitrate are dissolved separately in 10 ml water. After this we dissolved 701.25mg KOH in 25 ml water and then mixed the cadmium chloride and ammonium nitrate with KOH and raised the temperature of the bath at kept at 78^0C with constant stirring. For deposition , the desired area of pre-cleaned ITO substrate is suspended into the bath and mixed with 10 ml solution of thiourea to form CdS. The observed yellowish color indicates formation of nanocrytalline CdS . After about 40 minutes the suspended electrode is taken out, washed with distilled water and stored in a refrigerator.

2.2. Immobilization of ChOx onto nano-CdS/ITO electrode

Cholesterol oxidase is immobilized via physical adsorption onto the surface of prepared Nano-CdS/ITO electrodes. 30 µl of ChOx (1mg/dl) solution is spread onto the Nano-CdS/ITO electrode surface and the electrode is then left inside a humid chamber for about 4h for immobilization of ChOx on electrode surface and stored in 4°C when not in use. The ChOx binds with the CdS quantum dots via electrostatic interaction.

2.3. Instrumentations

Surface morphology and structural properties of electrode (Nano-CdS/ITO) and bioelectrode (ChOx/Nano-CdS/ITO) has been characterized using scanning electron microscopy (SEM, Zeiss EVO 40) and Fourier transform infra-red (FT-IR) spectroscopy (Perkin–Elmer ,model ''Spectrum BX'' using ATR accessory), UV–visible spectrophotometer (Model 2200DPCV, Phoenix) and X-ray diffraction(PAN alytical X'Pert PRO). Electrochemical investigations (CV and EIS) of the prepared electrodes have been carried out using an Autolab Potentiostat/Galvanostat (Eco Chemie, Netherlands) in a conventional three-electrode electrochemical cell consisting of Ag/AgCl as reference electrode and platinum foil as the counter electrode. Electrochemical impedance spectroscopy (EIS) studies have been performed in the frequency range, 0.01–10^5 Hz with amplitude of 5 mV in PBS solution.

3. Results and discussion

3.1. UV-visible and X-ray diffraction

The UV-visible absorption spectra of the Nano-CdS/ITO electrode is shown in Fig. 1(a). the characteristic absorption peak assigned to the first excitonic state of the CdS electrode is located at 450 nm with the broad range of 425-500 nm. The blue-shift is observed for the absorption band of Nano-CdS as compared to the 520nm of the bulk CdS. This indicates quantum confinement effect and presence of CdS nanoparticles/crystallites in the fabricated electrode[19]. The X-ray diffraction (XRD) pattern of the Nano-CdS/ITO electrode is shown in the Fig.1(b). The XRD pattern indicates polycrystalline nature of the Nano-CdS film deposited onto ITO glass. The CdS exists in two crystallographic forms, hexagonal (Wurtzite) and cubic (Zincblend). The presence of the peaks with 2θ values at 30.55°, 45.56° and 50.93° may be associated with cubic planes (111), (220) and (311) or with the hexagonal lattice planes (101), (110) and (200), respectively. The extra peaks seen at 35.80

and 60.6 are due to the ITO substrates. The average grain size (D) of the CdS is found to be as 12 nm, calculated using Scherrer formula $D=0.94\lambda/\beta \cos\theta$, where λ is the X-ray wavelength, β is the full width at half maximum and θ is the Bragg angle.

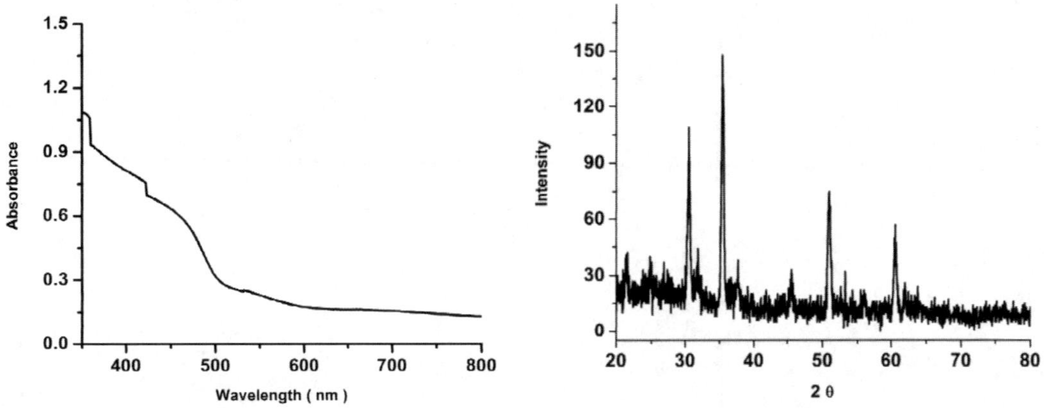

Figure 1. (a) UV-Visible spectra of Nano-CdS/ITO electrode & (b) XRD pattern of Nano-CdS/ITO.

3.2. Scanning electron microscopy studies

The scanning electron micrographs (SEM) of the Nano-CdS/ITO electrode shows the uniform and continuous crystallization during the film formation (Fig. 2(a)). The homogeneous crystalline granular and nanoporous structure of CdS with size distribution of 50- 100 nm can be seen on the ITO surface. The SEM image of the cholesterol oxidase (ChOx) immobilized onto the electrode surface Fig. 2(b) demonstrates the covered and packed granular structure indicating immobilization of ChOx via electrostatic interactions.

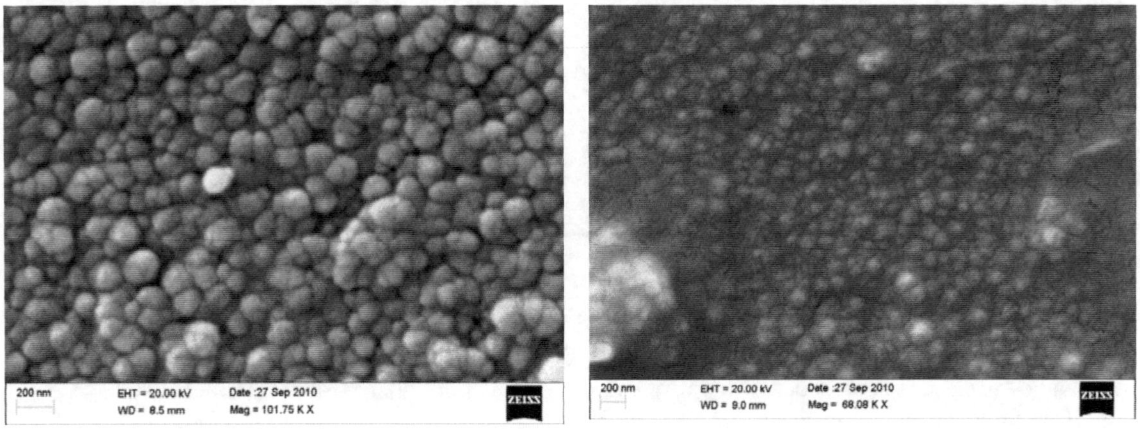

Figure 2. (a) SEM micrograph of Nano-CdS/ITO electrode & (b) ChOx/Nano-CdS/ITO bioelectrode.

3.3. Fourier transform infra-red spectroscopy studies

FT-IR spectra of Nano-CdS/ITO electrode is shown in Fig. 3(a). The peak seen at 671 cm^{-1} in the fingerprint region, corresponds to Cd-S vibrational band. The peaks observed at 1187 cm^{-1} , 1363 cm^{-1} and 1450 cm^{-1} are assigned to O-C=O (symmetric and asymmetric stretching band) due to strong physical absorption of CO_2 and H_2O on the electrode surface. The spectra [Fig. 3(b)] shows the broad peak at 1100 cm^{-1} confirms the immobilization of ChOx with the peak found at 1630 cm^{-1} is due to amide bonds showing the presence of ChOx[20].

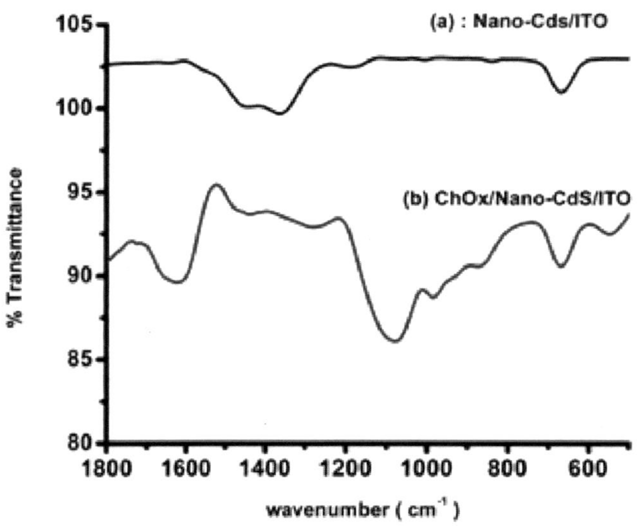

Figure 3. FT-IR spectra of (a) Nano-CdS/ITO electrode and (b) ChOx/Nano-CdS/ITO bioelectrode

3.4. Cyclic voltammetry studies

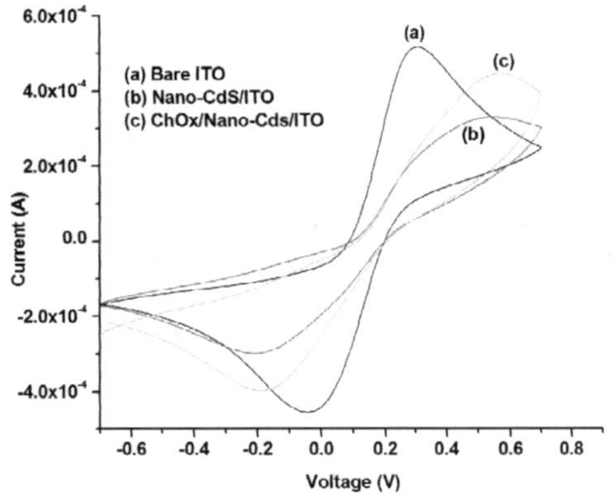

Figure 4. Cyclic Voltammetric (CV) of bare ITO (a), Nano-CdS/ITO electrode (b) and ChOx/Nano-CdS/ITO bioelectrode at scan rate 20 mV/s in PBS (50mM, pH 7.0, 0.9% NaCl) containing $[Fe(CN)_6]^{3-/4-}$ (5mM).

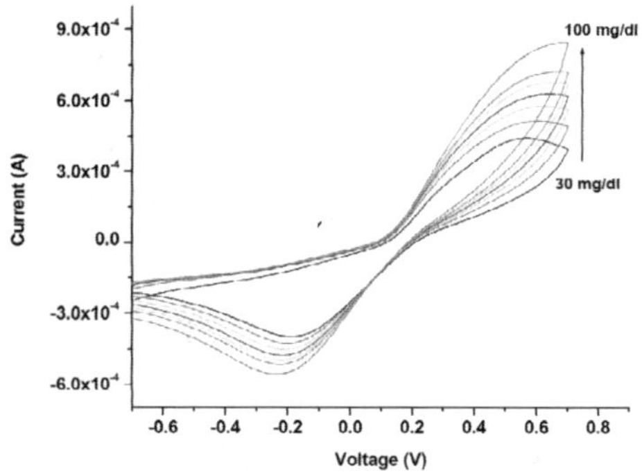

Figure 5. Cyclic Voltammetric (CV) of ChOx/Nano-CdS/ITO bioelectrode at different scan rate 30-100 mV/s in PBS containing $[Fe(CN)_6]^{3-/4-}$ (5mM).

Fig. 4 shows results of cyclic voltammetric (CV) studies conducted on bare ITO (a), Nano-CdS/ITO electrode (b) and ChOx/Nano-CdS/ITO bioelectrode (c) at scan rate 20 mV/s with potential range -7.0V to +7.0V in PBS (50mM, pH 7.0, 0.9% NaCl) containing $[Fe(CN)_6]^{3-/4-}$ (5mM). It can be seen that the magnitude of the peak current for Nano-CdS/ITO electrode (3.29×10^{-4}A) decreases as compared to bare ITO (5.1710^{-4}A) and the peak is shifted toward the higher potential. This is due to the semiconducting CdS that obstructs the electron transfer between the CdS QDs electrode and the solution. The peak current is found to be higher for ChOx/Nano-CdS/ITO bioelectrode (c) as compared to that of the Nano-CdS/ITO electrode (b) indicating that Nano-CdS quantum dots perhaps act as mediator resulting in enhanced electron transfer between enzyme and electrode.

Fig. 5 shows cyclic voltammogrammes obtained for ChOx/Nano-CdS/ITO bioelectrode in PBS containing $[Fe(CN)_6]^{3-/4-}$ recorded at different scan rates (30–100mVs^{-1}). It can be seen that the anodic potential shifts towards positive side and the cathodic peak potential (Fig.6A) shifts in the reverse direction. The peak-to-peak separation potential ΔE ~0.37 V increases with increasing scan rate resulting in uniform facile charge transfer kinetics. The magnitude of current of both anodic peak (I_a) and cathodic peak (I_c) observed for ChOx/Nano-CdS/ITO bioelectrode increases linearly with the scan rate indicating quasi reversible diffusion controlled behavior. The values of the slope, intercept and correlation coefficient given as

I_a [ChOx/Nano-CdS/ITO]= -1.3μA + 0.8μA (s/mV) × scan rate (mV/s) with r^2=.993 (Eq.1)

I_c[ChOx/Nano-CdS/ITO]= -0.002μA -0.35μA (s/mV) × scan rate (mV/s) with r^2=.998 (Eq.2)

3.5. Cyclic voltammetry studies

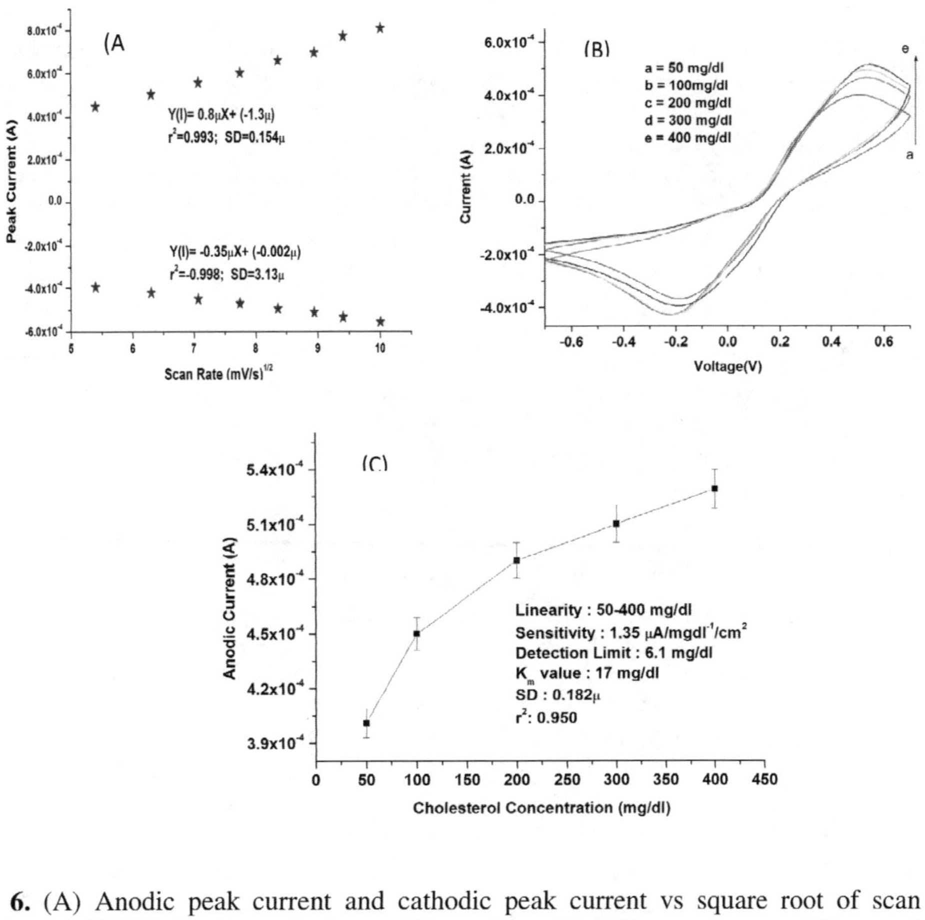

Figure 6. (A) Anodic peak current and cathodic peak current vs square root of scan rate of ChOx/Nano-CdS/ITO bioelectrode from CV studies, (B) CV response at different cholesterol concentration (50-400 mg/dl) at scan rate 20 mV/s in PBS (50mM, pH 7.0, 0.9% NaCl) containing $[Fe(CN)_6]^{3-/4-}$ (5mM) and (C) Sensor response curve plot between current vs cholesterol concentration.

The results of response studies of the ChOx/Nano-CdS/ITO bioelectrode for different cholesterol concentrations in phosphate buffer (50mM, pH 7.0, 0.9% NaCl) containing $[Fe(CN)_6]^{3-/4-}$ are shown in Fig. 6(B). The peak current of CV response increases with adding cholesterol indicating increase in the H_2O_2 concentration produced due to the interaction of ChOx with cholesterol during the biochemical reaction [Fig. 6C]. The ChOx/Nano-CdS/ITO bioelectrode shows linearity as 50–400 $mgdl^{-1}$ and the detection limit as 6.1 mg dl^{-1}. This biosensor shows high sensitivity as 1.35 $\mu A/mgdl^{-1}/cm^2$ with linear regression (r^2) 0.950. The Michaelis–Menten constant (K_m), that gives an indication of the enzyme-substrate kinetics, can be obtained from the Lineweaver–Burke plot . The K_m value for the ChOx/Nano-CdS/ITO bioelectrode has been calculated using the formula 3×SD/m, where SD is standered deviation and m is the slope of the curve and it is found to be 17.0 $mgdl^{-1}$ indicating high affinity for cholesterol.

3.6. Effect of interferents of the ChOx/Nano-CdS/ITO bioelectrode

The effect of interferents (lactic acid, ascorbic acid, uric acid, glucose and urea) on the response of this cholesterol sensor has been evaluated by adding the solution containing (1:1) ratio of cholesterol (100$mgdl^{-1}$) and interferents such as glucose (5 mM), ascorbic acid (0.05 mM), uric acid (0.1 mM), lactic acid (0.5mM) and urea (1 mM) (data not shown). The results indicate negligible effect of these interferents on the CV response of ChOx/Nano-CdS /ITO bioelectrode.

4. Conclusions

We have fabricated the nanocystalline CdS electrode using chemical bath deposition technique. The UV-Visible and X-ray diffraction studies confirms that the quantum confinement effect during the formation of CdS favors its deposition as homogeneous nanocrystals and helps in ChOx loading. FT-IR and SEM studies reveal the successful immobilization of ChOx onto Nano-CdS/ITOelectrode surface. The ChOx/Nano-CdS/ITO bioelectrode exhibits linear range of 50–400 mg/dl and improved sensitivity 1.35 $\mu A/mgdl^{-1}/cm^2$. It has been found that the CdS Qdots based bioelectrode shows the detection limit of 6.1 mg/dl and low K_m(0.45mM) value indicating strong enzyme-matrix affinity. Efforts should be made to control the shape and size of Nano-CdS and to utilize this electrode for estimation of total cholesterol, triglycerides and low density lipoproteins (LDL) etc.

Acknowledgements

The authors thank Director NPL, New Delhi India for providing the facilities. We also thank the Advanced Instrumentation and Research facility (AIRF), JNU, New Delhi for the XRD and SEM analysis. Hemant Dhyani, Saurabh Srivastava and Md. Azahar Ali thank CSIR-India for the award of Senior Research Fellowships.

References

[1] Hu M, Tian J, Lu H T, Weng L and Wang L H 2010 *Talanta* **82** 997-1002
[2] Frasco M F and Chaniotakis N 2009 *Sensors* **9** 7266-7286
[3] Jaiswal J K, Mattoussi H, Mauro J M and Simon S M 2003 *Nat. Biotechnol.* **21** 47-51
[4] Huang S, Xiao Q, Li R, Guan H L, Liu J, Liu X R, He Z K and Liu Y 2009 *Anal. Chim. Acta* 645-78
[5] Du P, Li H, Mei Z and Liu S 2009 *Bioelectrochemistry* **75** 37-43
[6] Liu Q, Lu X, Li L, Yao X and Li J 2007 *Biosens. & Bioelectron.* **22** 3203-3209
[7] Huang Y, Zhang W, Xiao H and Li G 2005 *Biosens. & Bioelectron.* **21** 817-821
[8] Zhao J, Chen G, Zhu L and Li G 2011 *Electro. Commu.* **13** 31-33
[9] Hansen J A, Wang J, Kawde A N, Xiang Y, Gothelf K V and Collin G 2006 *J. Am. Chem. Soc.* **128** 2228–2229
[10] Won Y H, Aboagye D, Jang H S, Jitianu A and Stanciu L A 2010 *J. Mater.Chem.* **20** 5030-5034
[11] Vastarella W and Nicastri R 2005 *Talanta* **66** 627-633
[12] Li Y, Kotzeva V P and Fray D J 2006 *Materials letters* **60** 2743-2746

[13] Prabhakar S, Suryanarayanan N, Rajasekar K, Srikanth S and Kathirvel D 2009 *Chalcogenide Letters* **6** 309-313
[14] Ramaiah K S, Pilkington R D, Hill A E, Tomlinson R D and Bhatnagar A K 2001 *Materials chemistry and physics* **68** 22-30
[15] Liu M, Shi G, Zhang L, Cheng Y and Jin L 2006 *Electrochem. Commu.* **8** 305-310
[16] Dhyani H, Ali A, Pandey M K, Malhotra B D and Sen P 2012 *J. Mater. Chem.* **22** 4970-4976
[17] Wang K, Liu Q, Guan Q M, Wu J, Li H N and Yan J J 2011 *Biosens. & Bioelectron.* **26** 2252-2257
[18] Flores A O, Perez P B, Rodr´ıguez R C and Oliva A I 2006 *Rev. Mex. Fis.* **52** 15–19
[19] Majumdar M, Chakraborty A K and Mallik B 2010 *Journal of Luminescence* **130** 1497-1503
[20] Solanki P, Kaushik A, Ansari A and Malhotra B D 2009 *Appl. Phys. Lett.* **94** 143901

Self-assembled monolayer based electrochemical nucleic acid sensor for *Vibrio cholerae* detection

Manoj K Patel [1, 2], Pratima R Solanki[1], Sachin Khandelwal[4], Ved V Agrawal[1], S G Ansari[2] and B D Malhotra[3]

[1]Department of Science and Technology Centre on Biomolecular Electronics, Biomedical Instrumentation Section, National Physical Laboratory (CSIR), Dr. K. S. Krishnan Marg, New Delhi -110 012, India

[2]Centre for Interdisciplinary Research in Basic Sciences, Jamia Millia Islamia, New Delhi 10 025, India

[3]Department of Biotechnology, Delhi Technical University, Shahbad Daulatpur, Main Bawana Road, Delhi 110 042, India

[4]Microbiology Division, National Centre for Disease Control, 22, Sham NathMarg, Civil Lines, New Delhi 110 054, India

E-mail: bansi.malhotra@gmail.com (B D Malhotara)

Abstract. Nucleic acid sensor has been fabricated by immobilization of thiolated (5' end) single stranded deoxyribonucleic acid probe (ssDNA-SH) onto gold (Au) coated glass electrode for *Vibriocholerae*detection. This ssDNA-SH/Au bioelectrode characterized using atomic force microscopy (AFM),Fourier transforms infrared spectroscopy (FT-IR) and electrochemical technique, has been used for hybridization detection of genomic DNA (dsDNA/Au). This ssDNA-SH/Au bioelectrode can specifically detect up to 100- 500 ng/μL genomic DNA of *Vibriocholeare* within 60 s of hybridization time at 25°C by cyclic voltammetry (CV) using methylene blue (MB) as electro-active DNA hybridization indicator. The value of sensitivity of the dsDNA/Au electrode has been determined as 0.027μA/ng cm^{-2} with regression coefficient as 0.978. This DNA bioelectrode is stable for about 4 months when stored at 4°C.

1. Introduction

Cholera is endemic in southern Asia and parts of Africa and Latin America, where seasonal outbreaks occur widely and are particularly associated with poverty and poor sanitation. [1-2]. It is an acute intestinal infection caused by ingestion of food or water contaminated with the bacterium *Vibrio cholerae,* which is a Gram-negative bacterium that produces cholera toxin (CT), an enterotoxin, whoseaction on the mucosalepithelium lining of the small intestine is responsible for the disease's most salient characteristic, exhaustive diarrhea [3-4]. Most of the strains of *V. cholerae* are non-virulent but O1 and O139 is responsible for disease[5-6]. As compared to O139 antigen, O1 antigen has been found in majority of patients affected by *V. cholerae*whichcurved-rod shaped and oxidase-positive bacterium [7].

[3] To whom any correspondence should be addressed.

Many biomolecules such as gangliosides, antibodies, cholera toxin and somatic antigens (O1 and O139) have been used for *V. cholerae* detection. CT has become a very important compound in detecting bioterrorism, and there has been increasing interest in the development of rapid and sensitive methods for the determination of CT [8]. The conventional methods for *V. cholerae* detection include enzyme linked immunosorbent assay/ immunological detection kits, southern hybridization and polymerase chain reaction [6, 9-11]. However, all these methods are time-consuming, expensive and require expertise [12]. There is increased interest towards the application of biosensors for fast, low-cost, sensitive and reliable detection of *V. cholerae*. The deoxyribonucleic acid (DNA) biosensors are potentially useful for clinical applications to determine the pathogenic bacteria [13].

Biological macromolecules, such as nucleic acids, both (DNA) and ribonucleic acid (RNA) are known to play an important role towards the rapid development of molecular genomics, biotechnology and medical diagnostics [14]. Nucleic acid based biosensors based on the hybridization technique are gaining more attention over traditional diagnostic methods as they are faster, simpler and economical [15-17]. Several techniques have been used for DNA detection including fluorescence detection, mass spectrometry, electrochemical, surface plasmon resonance spectroscopy and quartz crystal microbalance [18]. Besides this, an electrochemical detection technique has advantages like simplicity, speed, low-cost, high sensitivity and specificity as compared to molecular detection approaches [19].

Self-assembled monolayers (SAMs) have been used in electro analytical chemistry due to their uniqueness in terms of well-ordered structure, ease of preparation, flexibility, reproducibility, organization, molecular control and low non-faradaic currents etc. [20-23]. The thiol groups chemisorbed onto the gold surface via the formation of a gold thiol bond [24-25] to produce a densely packed, highly ordered monolayer. The self-assembly process is mediated through non-specific adsorption, electrostatic interactions, covalent bonding and specific interactions. The SAMs are spontaneously formed by the adsorption/binding of organic molecules from solution or the gas phase onto a solid-state surface [26]. Zhao et al. have utilized hydroxyl-terminated SAM for covalent immobilization of dsDNA on gold surface [27]. SAM based [28] surface plasmon [29] and piezoelectric [30] immunosensors have been used for detection of *Salmonella Typhimurium, Leukemia* and *Escherichia coli* (*E. coli*). Prabhakar et al. have reported 5' thiol end labeled DNA probe for detection of *M. tuberculosis* [31].DNA biosensor based on electrochemical hybridization using methylene blue as indicator has been reported for detection of Chronic Myelogenous Leukemia, Human apolipoprotein E, Bar Gene, *E.coli*, etc. using glassy carbon electrode, graphite screen printed electrode, stearic-acid-modified carbon paste electrode, polyaniline, respectively [32-34].

We report the results of the immobilization of 5'-thiol end labeled DNA probe (23 bases) and hybridization with genomic DNA of *Vibrio cholerae* on Au bare surface were studied using Methylene blue (MB) as an electroactive indicator.

2. Materials and methods

2.1. Chemicals and reagents

Tris (hydroxymethyl) amino methane, ethylene diamine tetra acetic acid (EDTA), sodium dihydrogen ortho-phosphate and *di*-sodium hydrogen orthophosphate were procured from Qualigens, India and Methylene blue (MB) was procured from Sigma-Aldrich, USA. HPLC purified 23 bases thiolated modified oligomer probe (ssDNA-SH) 5'-HS - GCA TAT GCA AAT GGA ACA CCT CA- 3' were procured from GenxBio Health Sciences Pvt. Ltd, India. All reagents were prepared in de-ionized water (Milli Q 10 TS) and solutions were autoclaved prior to being use. DNA solutions were prepared in TE buffer (10 mMTris, 1mM EDTA, pH 8.0). All Bacterial isolates and genomic DNA sample were obtained from patients at the National Centre Diseases Control (NCDC), New Delhi, India. Bacterial isolates from feces were cultured on Thiosulfate Citrate Bile Salts Sucrose Agar (*TCBS* Media) and grown at 35°C for 24-48 h and the combination of alkaline peptone water.

2.2. Fabrication of ssDNA-SH/Au bioelectrode

Gold coated glass electrodes were first cleaned with piranha solution (H_2SO_4 and H_2O_2 in 7:3 ratio) for 5 min. After dipping in 70 % ethanol for 2-3 min, the electrode was washed with water and dried at room temperature. The thiolated DNA probe (ssDNA-SH) 6.6 ng/10 μl was then immobilized onto gold electrode area (0.5 cm^2) and kept for 24 h at 25^0C. The unbound probe was removed by several washings with TE buffer and dried. Genomic DNA(dsDNA) was hybridized (after denaturation at 95^0C) for 5 min with immobilized probe onto gold electrode surface. After hybridization the electrode was washed thrice with buffer to remove the unbound ssDNA. The schematic shows the immobilization and hybridization procedure in Scheme 1.

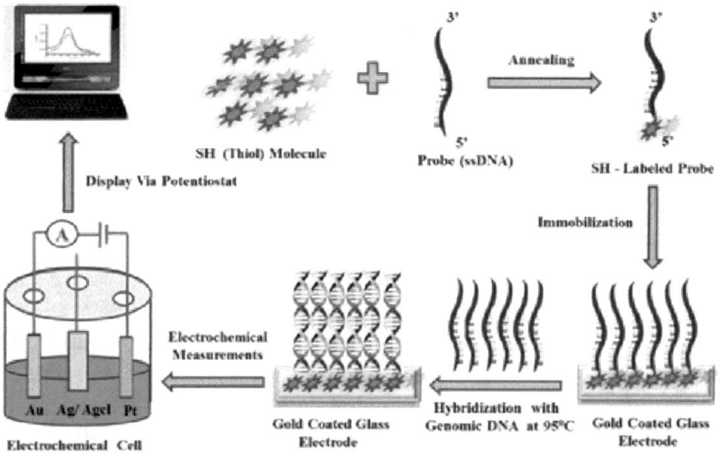

Scheme 1. Schematic representation shows the synthesis of SH labeled probe from SH molecule and hybridization of genomic DNA (dsDNA) with immobilized probe.

2.3. Sensor characterization

The surface morphology of ssDNA-SH/Au electrode and dsDNA/Au electrode were delineated using atomic force microscopy (VEECO, USA) in non-contact mode. Electrochemical experiments were carried out with a Electrochemical analyzer (Model AUT-84275, Auto lab) using three-electrode system containing gold coated glass electrode as working, Ag/AgCl as reference and platinum as counter electrode. MB (40 μM) was used as redox indicator in phosphate buffer saline (PBS, 0.05 M, pH 7.0, 0.9% NaCl). The cyclic voltammetric experiments were performed at 20 mV/s in PBS buffer at 25^0 C. Contact angle (Dataphysics, Model OCA15EC) measurement was done by sessile drop method before and after the immobilization/hybridization of DNA.

3. Results and discussion

3.1. AFM studies

Figure 1A. shows results of morphological studies obtained using AFM of bare Au electrode (image a), ssDNA-SH/Au electrode (image b) and dsDNA/Au [hybridization with genomic DNA(image c)] after denaturation at 95°C. Image (a) shows the nonporous morphology with uniform distribution of granular structure with roughness 6.68 nm (Fig.1Aa) and after the probe immobilization the roughness increases (18.4 nm) due to formation of well aligned self- assembled monolayer of thiolated ssDNA fabrication onto Au surface (Scheme 1) that filled the all the pores present onto Au surface. However, the surface morphology after the hybridization with genomic DNA (image c) has been observed as 16.3 nm that indicate that genomic DNA hybridized with the DNA probe present onto gold surface.

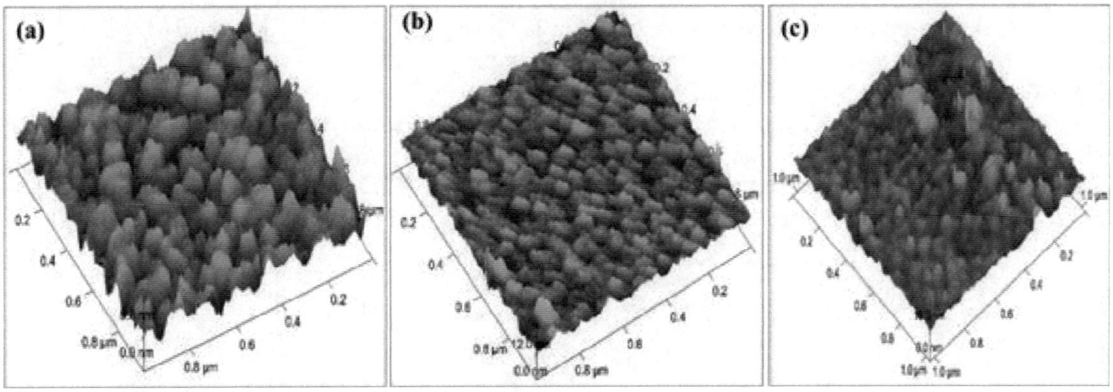

Figure 1A. AFM micrographs of bare Au electrode (image a), (image b) after immobilization with thiollabeled probe (ssDNA - SH/Au) and (image c) after hybridization with genomic DNA(dsDNA-SH/Au) onto Au surface.

3.2. Contact angle measurements

Contact angle measurements (Sessile drop method) have been carried out to investigate the formation of self-assembled monolayer onto Au surface. Contact angle studies have been carried out using deionized distilled water to characterize the immobilization of probe (ssDNA) and after hybridization with genomic DNA (dsDNA) onto the gold surface (Fig.1B). The change in the value of the contact angle reveals the hydrophobic/hydrophilic character of the surface due to immobilization of thiolated probe and hybridization with genomic DNA molecules. The contact angle value of the bare gold film (image a) is found as 76.46^o and decreases to 55.56^{o}after the immobilization of thiolated ssDNA probe (image b) and further decreases to 55.33^{o}after DNA hybridization (image c). The decrease in the contact angle values can be attributed to the presence of hydrophilic group (OH) in DNA probe.These results indicate successful binding of the nucleic acid (probe/DNA) onto Au electrode surface.

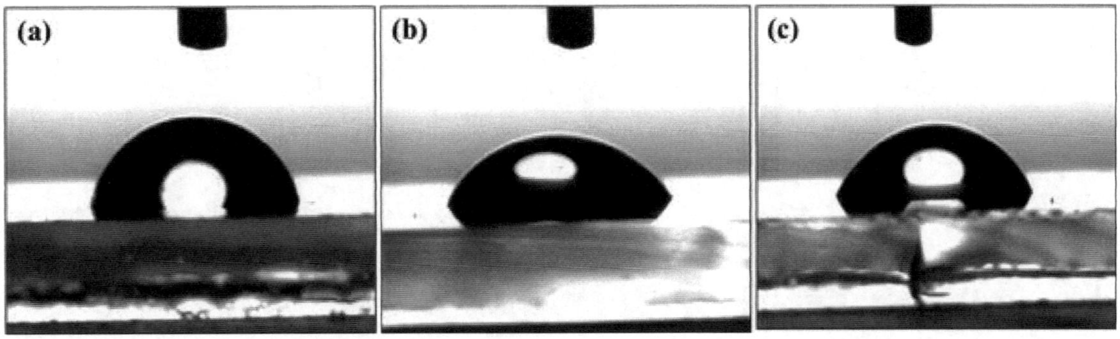

Figure 1B. Contact angle measurements of (a) bare Au electrode (76.46^o) (b) immobilization ssDNA - SH/Au (55.56^o) and (c) dsDNA-SH/Au (55.33^o) on gold coated glass electrode.

3.3. FT-IR studies

The FT-IR spectra of (a) ssDNA-SH/Au bioelectrode (b) dsDNA/Au exhibits peak at 891, 1047, 1215, 1392, 1461, 1455, 1515, 1548, 1650, 1903, 2066, 3261 cm^{-1} corresponding to the functional groups of DNA (Fig. 2). The peaks found at 891 to 1215 cm^{-1} correspond to asymmetric and symmetric PO_4^- groups of the phosphodiesterdeoxyribose backbone. The peaks seen at 1455, 1515, 1548 and 1650cm^{-1} correspond to cytosine, adenine, thymine and guanine respectively. A FT-IR spectrum after hybridization with both amplified and genomic DNA shows similar peaks of different bases with increased peak intensity. The peaks seen at 1515, 1548 and 3261 cm^{-1} are attributed to N-H bending and stretching of DNA bases purine and pyrimidine ring [35]. The 1515 - 1650 cm^{-1} peaks correspond to C=O, C=N, C=C stretching and exocyclic $-NH_2$ bending vibrations in the DNA bases. These results indicate the immobilization of DNA probe onto Au surface and hybridization with genomic DNA.

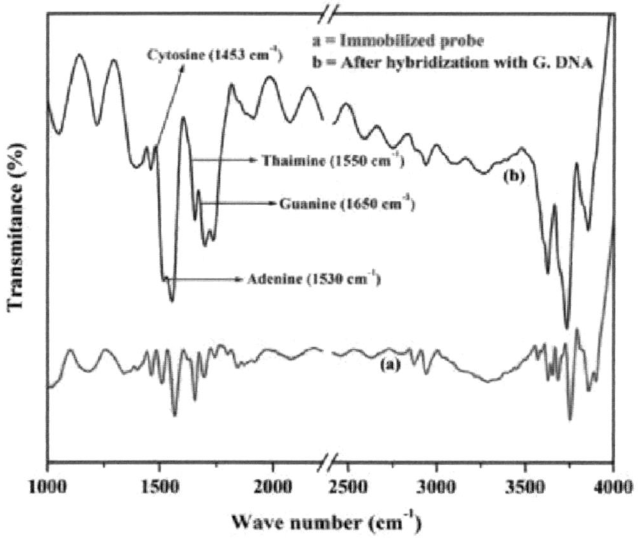

Figure 2. FT-IR spectra of (a) ssDNA - SH/Au and (b) dsDNA-SH/Au of *Vibrio cholerae.*

3.4. Electrochemical impedance studies

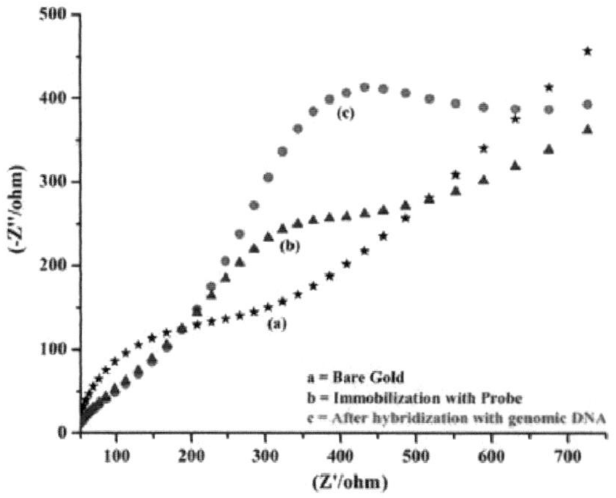

Figure 3. Impedance spectra of (a) bare Au electrode (b) ssDNA - SH/Au and (c) dsDNA-SH/Au of O1 gene in PBS containing 5 mM $[Fe(CN)_6]^{3-4-}$.

Figure 3 exhibits the Faradic impedance spectra in the frequency range 0.01-10^5 Hz, presented by the Nyquist plots of (a) bare gold (b) ssDNA-SH/Au and (c) dsDNA/Au in PBS containing 5 mM $[Fe(CN)_6]^{3-/4-}$. The value of *R*ct, charge-transfer resistance, obtained as 3.75 x $10^2\ \Omega$ (curve a) for bare gold is lower than that of ssDNA-SH/Au electrode. This smaller semicircle diameter indicates faster electron-transfer kinetics of $[Fe\ (CN)_6]^{3-/4-}$ on bare Au electrode (curve a). However, after the immobilization of ssDNA-SH (curve b), Rct value increases (1.09 x$10^3\ \Omega$) due to presence of electro-negative phosphate skeletons that perhaps prevent $[Fe(CN)_6]^{3-/4-}$ions from reaching the electrode surface for electron transfer during redox reaction. This reveals that ssDNA-SH has been immobilized onto Au surface. Further, increase in the *R*ct value after hybridization of probe with the genomic DNA (6.74 x $10^2\Omega$; curve c) can be attributed to the increased concentration of negatively charged backbone of DNA onto the surface of the electrode that may cause repulsion between the similarly charged redox couple of the system. The increase in the Rct, value indicates duplex form of DNA indicating immobilization of ssDNA hybridization with genomic DNA.

3.5. Cyclic voltammometric studies

Figure 4A shows results the cyclic voltammetric (CV) carried out for (a) bare gold (b) ssDNA/Au electrode and (c) dsDNA-SH/Au (500 ng) electrode in PBS containing methylene blue (40 μM). Methylene blue (MB) is an intercalating organic compound has been used for electrochemical DNA detection as their response changes upon DNA hybridization. It can be seen that the oxidation peak current for bare gold electrode is higher (6.80 μA, curve a) than that of ssDNA-SH/Au electrode (0.74 μA, curve b), indicating immobilization of the DNA probe onto Au surface. The higher oxidation peak current of bare Au surface as compared to that of the ssDNA-SH/Au electrode, is attributed to the enhanced flow of electrons between the gold electrode and MB. On the other hand, the oxidation peak current of dsDNA-SH/Au bioelectrode (4.64 μA, curve c) increases as compared to that of the ssDNA-SH/Au bioelectrode (curve b). It appears that dsDNA-modified surface is more favorable for MB molecules to undergo redox reaction than bare gold surface, indicating a pre-concentration of MB on the DNA modified surface due to intercalation. The sharp cathodic peak also indicates that MB underwent redox through the intercalated state [36]. Moreover, the highest MB oxidation signal was obtained at the electrode hybridization with genomic DNA sequence (curve c), indicating that the

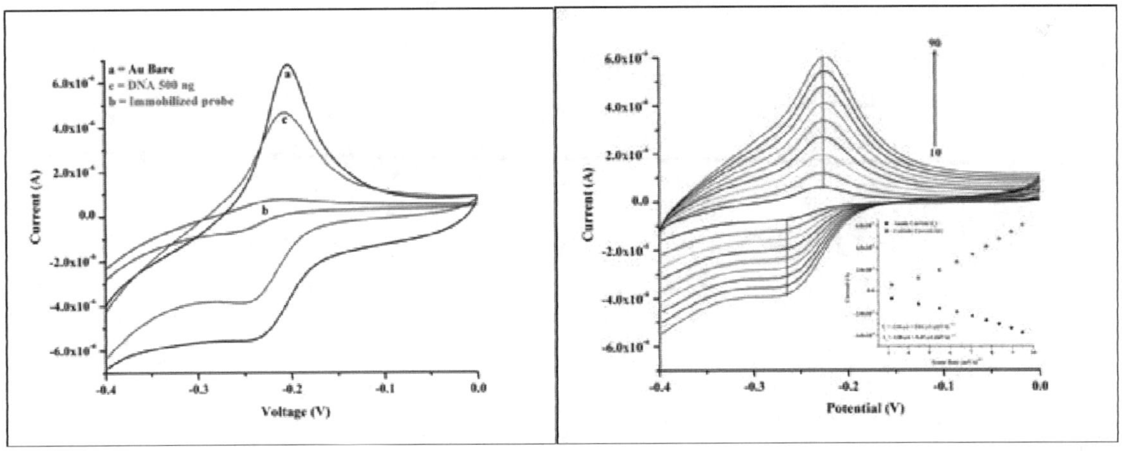

Figure 4A Cyclic voltammogram of (a) bare Au electrode (c) dsDNA-SH/Au bioelectrode and (b) ssDNA-SH/Au electrode 500 ng at 20 Vs^{-1} in PBS containing 40 μM MB and (B)Cyclic voltammogram of ssDNA-SH/Au bioelectrode on increasing scan rate from 10 to 90 at 20 Vs^{-1} in PBS (50 Mm, pH 7 containing 0.9 % NaCl) in 40 μM MB.

largest attachment of MB occurred at this electrode because of the strong affinity of MB with the formed dsDNA. The peak current (curve b) of MB decreases after the ssDNA probe [37]. This result can be perhaps be due to stronger association of MB molecules with unpaired nitrogenous bases of ssDNA-SH/Au bioelectrode (curve c), especially with guanine bases as compared to dsDNA. The ssDNA-SH/Au bioelectrode consists of the limited number of unhybridized nitrogenous bases that interact with MB, while genomic DNA comprises of prolonged unhybridized ssDNA sequence consisting of uncountable nitrogenous bases, resulting in increased MB molecules interaction. Thus oxidation current is found to be increased in case dsDNA-SH/Au bioelectrode as compared to that of the ssDNA-SH/Au electrode. The magnitude of the peak current of ssDNA-SH/Au electrode has been found to increases with increased scan rate (Fig. 4B), indicating slow electron-transfer kinetics.

3.6. Electrochemical response studies

The response studies have been carried out using cyclic voltammograms at scan rate of 20 Vs^{-1} in PBS containing 40 μM MB (Fig. 5). The ssDNA/Au bioelectrode has been hybridized with different concentrations (100-500 ng) of genomic DNA for 5 minutes at room temperature (25°C). It has been observed that the magnitude of anodic peak current of MB increases with increase in the concentration of double stranded DNA (genomic DNA). The inset in Fig. 5 shows the curve between the magnitudes of peak current increases with concentration of genomic DNA. It appears that positively charged MB molecules are electrostatically attached to negatively charged phosphate backbone of double stranded DNA or get intercalated after hybridization. Moreover, the observed enhanced MB peak current with increased concentration of genomic DNA due to presence of more free guanine bases (unhybridized bases) of genomic DNA indicates increased interaction of MB with bases resulting in increased peak current [38]. It has been observed that increase in the MB peak with respect to genomic DNA concentration follows Eq. (1).

$$I_{\text{hybridized DNA}} = 96.8 \times 10^{-2} + 6.9 \times 10^{-3} \times [\text{genomic DNA conc.}] \text{.......... (Eq.1)}$$

The specificity of the ssDNA/Au electrode has been monitored with the non-complementary DNA, resulting in no change in current as compared to that of ssDNA probe, indicating that no

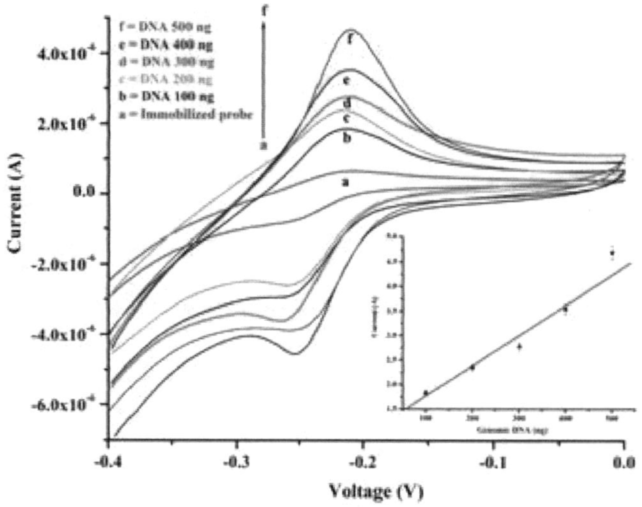

Figure 5. Response studies of ssDNA-SH/Au after hybridization with different concentration of genomic DNA (100-500) at 20 Vs^{-1} in PBS (50 Mm, pH 7 containing 0.9 % NaCl) in 40 μM MB and inset shows the linear graph between current and concentration of genomic DNA.

hybridization occurs. The hybridization time of probe with genomic DNAhas been obtained as about 5 min. The sensitivity of dsDNA/Au electrode has been obtained as 0.027μA/ng cm^{-2} with 0.978 regression coefficients (R). The ssDNA/Au electrode can detect *Vibrio cholerae* in the range of 10-60 ng/μl. The stability of ssDNA/Au electrode is obtained to be about 4 months when stored at 4°C.Table 1 shows characteristics of the ssDNA-SH/Au bioelectrode for cholera detection along with those reported in the literature.

Table 1. The characteristics of ssDNA-SH/Au bioelectrode for cholera detection along with those reported in the literature.

Sl. No.	Surface/matrix	Transducer	Sensitivity	Detection limit	Linearity	Reference
1	Gold transducer surface	Quartz crystal microbalance	-------- -	10^5 cells/ ml	10 MHz AT	[39]
2	Gold-coated AFM microcantilevers	Microcantilever-based biosensor	146.5 pg/Hz	1×10^3 CFU/ml	1×10^{-3} to 1×10^7 CFU/ml	[40]
3	Polytyramine-modified gold electrode	Immunosensor	9×10^{-20} M or 0.09 aM	9×10−20 M or 0.09 aM	0.1 aM and10pM	[41]
4	Au electrode	DNA biosensor	0.027μA/ng cm^{-2}	100 ng	100-500 ng	Present Work

4. Conclusions

A self-assembled monolayer based electrochemical DNA biosensor has been fabricated by immobilizing thiolated probe (ssDNA-SH/Au) onto gold electrode surface and it's hybridized with genomic DNA (dsDNA/Au) for detection of *Vibrio cholerae*. It has been concluded that ssDNA-SH/Au exhibits higher sensitivity (0.027μA/ng cm^{-2}) with regression coefficient as 0.978 (R). The hybridization time of probe with genomic (dsDNA) at Au electrodes is 5 min with response time 60s. This electrochemical DNA sensor is highly specific for the detection of *Vibrio cholerae* in clinical samples. Efforts are being made to utilize this DNA probe for diagnosis of Cholera directly using patient's samples.

Acknowledgements

Authors thank Director, NPL, New Delhi, India for the facilities. MKP is grateful to the Council of Scientific and Industrial Research (CSIR), India for the award of Senior Research Fellowship (SRF). Authors are thankful to Md.Azahar Ali for suggestion and support and thankful to Mr.Sandeep Singh for AFM measurements.

References

[1] Alam M, Hasan N A, Sultana M, Balakrish G, Sadique N A, Faruque A. S. G, Endtz H P, SackR B, Huq A, Colwell R R, Izumiya H, Morita M, Watanabe H and Cravioto A 2010 *J. Clin.Microbiol.* **48** 3918-3922

[2] Faruque S M, Albert M J and Mekalanos J J 1998 *Microbiol. Molecul. Biol. Rev.* **68** 1301-1314

[3] Kelly P 2011 *Medicine* **39** 201-206

[4 Fournier J M and Quilici M L 2007 *Presse Medicale* **36** 727-739
[5] Louis V, Estelle R, Choopun N, Rivera I N G, Gangle B, Jiang S C, Rubin A, Patz J A, Hug A and Colwell R R 2003 *Appl. Environmen. Microbiol.* **69** 2773-2785
[6] Gubala A J 2006 *J. Microbiological Meth.* **65** 278-293
[7] Jyoung J Y, Hong S, Lee W and Choi J W 2006 *Biosens. Bioelectron.* **21** 2315-2319
[8] Nikoleli G P, Nikolelis D P and Tzamtzisa N 2011 *Electroanal.* **23** 2182-2187
[9] Martinez G A, Ambrosio J, Gutierrez C L and Flisser A 2001 *Clinic. Diagno. Lab. Immunol.* **28** 768-771
[10] Rivera I N G, Lipp E K, Gil A, Choopun N, Huq A and Colwell R R 2003 *Environ. Microbiol.* 599-606
[11] Ligler F S, Taitt C R, Shriver L L C, Sapsford K E, Shubin Y and Golden J P 2003 *Anal. Bioanal. Chem.* **377** 469-477
[12] Yamazaki W, Seto K, Taguchi M, Ishibashi M and Inoue K 2008 *BMC Microbiology* **8** 94
[13] Belluzo M S, Ribone M E and Lagier C M 2008 *Sensors* **8** 1366-1399
[14[Ciminska M G 2006 *Microbial Cell Fact.* **5** 9
[15] Patel M K, Solanki P R, Kumar A, Khare S, Gupta S and Malhotra B D 2010 *Biosens. Bioelectron.* **25** 2586-2591
[16] Souteyrand E, Cloarec J P, Martin J R, Wilson C, Lawrence I, Michelson S and Lawrence M F 1997 *J. Phys. Chem. B.* **101** 2980-2985
[17 Berney H, West J, Haefele E, Alderman J, Lane W and Collins J K 2000 *Sens. Actua. B.* **68** 100-108
[18] Arora K, Chand S and Malhotra B D 2006 *Anal. Chim. Acta* **568** 259-274
[19] Wei F, Lillehoj P B and Ho C M 2010 *Pediatr Res.* **67** 458-468
[20] Matharu Z, Solanki P R, Gupta V and Malhotra B D 2012 *Analyst* **137** 747-753
[21] Arya S K, Solanki P R, Datta M and Malhotra B D 2009 *Biosens.Bioelectron.* **24** 2810-2817
[22] Gooding J J, Mearns F, Yang W and Liu J 2003 *Electroanal.***15** 81-96
[23] Cabrita J F, Abrantes L M and Viana A S 2005 *Electrochim. Acta* **50** 2117- 2124
[24] Gooding J J, Pugliano L, Hibbert D B and Erokhin P 2000 *Electrochem. Commun.* **2** 217- 221
[25] Li Y, Huang J, Mciver R T and Hemminger J C 1992 *J. Am. Chem. Soc.* **114** 2428-2432
[26] Kim D C and Kang D J 2008 *Sensors* **8** 6605-664
[27] Zhao Y D, Pang D W, Hu S, Wang Z L, Cheng J K and Dai H P 1999 *Talanta* **49** 751-756
[28] Subramanian A, Irudayaraj J and Ryan T 2006 *Biosens. Bioelectron.* **21**998-1006
[29] Oh B K, Kim Y K, Park KW, Lee W H and Choi J W 2004 *Biosens. Bioelectron.* **19** 1497-1504
[30] Su X L and Li Y 2004 *Biosens. Bioelectron.* **19** 563-574
[31] Prabhakar N, Arora K, Arya S K, Solanki P R, Iwamoto M, Singh H and Malhotra B D 2008 *Analyst* **133** 1587-1592.
[32] Lin X, Wu P, Chena W, Zhang Y and Xia X 2007 *Talanta* **72** 468-471
[33] Marrazza G, Chiti G, Mascini M and Anichini M 2000 *Clin. Chem.* **46** 31-37
[34] Arora K, Prabhakar N, Chand S and Malhotra B D 2007 *Anal. Chem.* **79** 6152-6158
[35] Patel M K, Solanki P R, Seth S, Gupta S, Khare S, Kumar A and Malhotra B D 2009 *Electrochem. Commun.* **11** 969-973
[36] Yau H C, Chan H L and Yang M 2003 *Biosens. Bioelectron.* **18** 873-879
[37] Gao H, Qi X, Chen Y and Sun W 2011 *Anal. Chim. Act.* 704 733-738
[38] Solanki P R, Patel M K, Kaushik A, Pandey M K, Kotnala R K and Malhotra B D 2011 *Electroanal.* **23** 2699-2708
[39] Chen H, Zheng Y, Jiang J H, Wu H L,Shen G L and Yu R Q 2008 *Biosens. Bioelectron.* **24** 684-689
[40] Sungkanak U, Sappat A, Wisitsoraat A, Promptmasa C, Tuantranont A 2010 *Biosens. Bioelectron.* **26** 784-789.
[41] Loyprasart S, Hedstroma M, Thavarungkul P, Kanatharana P, Mattiasson 2010 *Biosens. Bioelectron.* **25** 1977-1983.

Direct probing of carrier motion and interfacial phenomena in organic films and devices by optical second harmonic generation and Maxwell-displacement-current

M Iwamoto[1]
Department of Physical Electronics, Tokyo Institute of Technology,
2-12-1 O-okayama, Meguro-ku, Tokyo 152-8552, Japan

E-mail: iwamoto@pe.titech.ac.jp

Abstract. Probing and modelling of dynamical motions of carriers in organic materials is a fundamental research subject in science and electronics. According to the Maxwell's electromagnetic field theory, electrons and dipoles are source of electric fields. As a result organic materials surrounding electrons and dipoles are polarized. By probing the dielectric polarization, we can visualize carrier motion in organic materials. An optical method based on electric-field-induced optical second harmonic generation (EFISHG) is available for directly probing the dynamical electron (and hole) transport as the migration of electric field. On the other hand, an electrical method based on the Maxwell-Displacement current generation can probe rotational dipolar motion in organic films. In this proceeding, basic concept for probing carrier motions, e.g. translational motion of electrons and rotational motion of dipoles, by paying attention to dielectric polarization phenomena is discussed. Finally, it is concluded that experiments and analyses based on dielectrics physics is a very effective way for analyzing carrier behaviours in organic films and devices.

1. Introduction

Probing and modelling of polarization phenomena originating from dipoles and electrons is a fundamental research subject in electronics and materials science. According to the Maxwell's electromagnetic field theory, the total current flowing across organic materials is the sum of the conduction current and Maxwell's displacement current (MDC). The conduction current is generated when electrons and holes are conveyed under the external electric field produced by applying potentials to electrodes. In contrast, the MDC is generated when electric flux originating from electrons, holes and dipoles changes with time. By probing the change of electric flux with time as MDC, we could study the transient-state carrier motions that include rotational dipolar motion in organic materials [1,2].

In the MDC experiment, organic materials are sandwiched between two electrodes that are connected each other through an electrometer. On applying an external stimulus (for example, light, pressure, heat or external electric field) to organic materials, a current pulse is generated and it is recorded as the trace of change of induced charge on the electrodes with time. Note that we do not need to contact the two electrodes to organic materials for probing rotational motion of dipoles in the MDC measurement, where an air-gap between the electrode and organic materials works as a good

[1] To whom any correspondence should be addressed.

Published under licence by IOP Publishing Ltd

electrical insulator and eliminates leakage currents. Accordingly we can probe dipolar motions in organic materials as the MDC. Further, by coupling the MDC experimental system with Brewster-Angle microscope (BAM), we could visualize surface dielectric polarization phenomena in monlayers on the water surface [3].

On the other hand, a long-range carrier-(electron and hole) transit through organic materials is fundamental, and time-of-flight (TOF) measurement has been used for the probing. On applying a laser pulse to organic materials, induced charge on electrodes is monitored along with carrier-transit by using an electrometer. Consequently TOF is basically the MDC measurement, but we need to install a system of organic materials that contact with electrodes. For the MDC experiment used for probing dipolar motion in monolayer at the air-water interface, the location of dipoles is known *a-priori*. It is of course on the water surface. On the other hand, we have no idea on the location of moving carriers in organic materials, *a-priori*. Hence for the TOF measurement, an elegant mathematical approach for analyzing the current pulse is needed. However, the method leads to a puzzling situation in which there are many possible solutions that can reproduce an observed current pulse. This means it is difficult to probe actual carrier motion in organic materials by the TOF method. The only way to overcome this situation is to directly probe moving carriers in organic materials. Although it is a difficult task to probe carriers themselves by using microscope, nanoscope and others, the Gauss's law in the Maxwell's electromagnetic field theory suggests us an insightful principle for probing the carrier motion in organic materials. Thai is by probing the propagation of the electric field diverging from moving electrons and holes, the carrier motion could be visualized if the organic material surrounding the moving electrons and holes is polarized by the electric field arising from the moving carriers. Time-resolved microscopic optical second harmonic generation (TRM-SHG) measurement is a technique that enables us to probe the propagation of the polarization induced from moving electrons and holes [4]. Further, the use of CCD camera visualises the motion [5].

2. Probing of dipolar motion by MDC measurement

On account of symmetry breaking, monolayers are spontaneously polarized at the interface, where orientational ordering of the constituent dipoles makes a main contribution. The orientational distribution of rod-like molecules is expressed using a set of order parameters defined as S_n ($\equiv\langle P_n(\cos\theta\rangle$) (n=1, 2, 3, …), where $P_n(\cos\theta)$ is the Legendre polynomial of the n-th rank, θ is a tilt angle from the normal direction to the surface, and $\langle\ \rangle$ represents thermodynamic average. S_1 and S_3 express the non-symmetric property of monolayers. Spontaneous and nonlinear polarizations of monolayers are thus described using these parameters. On the other hand, S_2 specifies the symmetric property, and linear polarization is expressed using this parameter. The liquid crystalline property of bulk nematic LCs is generally characterized using the parameter S_2. Interestingly, a set of order parameters S_1, S_2 and S_3 are non-zero for monolayers on the water surface. In other words, non-zero S_1 and S_3 have relation with the presence of spontaneous polarization. This means that electrostatic energy stored in domains on a water surface governs domain shapes. Experimentally, by coupling the MDC experimental system with Brewster Angle Microscope (BAM), spontaneous polarization is probed with the order parameter S_1. At the same time, the BAM image visualizes domain shapes on the water surface

In more detail, an experimental arrangement of the MDC with a BAM observation system has been constructed, where a suspended electrode is used to probe electric field generated from monolayer on water surface. Induced charge Q_S on this suspended electrode is proportion to the spontaneous polarization of the monolayer P_0. Here P_0 is thus proportion to μS_1, where μ is permanent dipole moment of the constituent molecule of monolayer, and S_1 is the orientational order parameter. Monolayer compression causes the change of Q_S, and MDC current pulse is thus generated. This means that MDC probes the orientational change of molecules on the water surface, and the S_1 is evaluated. Along with the monolayer compression, the electrostatic energy is stored in monolayers. Hence domain shape of monolayer at the air-water interface should be dependent on the electrostatic energy of monolayers. Minimizing the free energy of monolayers, the shape equation describing domain shapes is derived [6], where the Maxwell-stress term (2nd and 3rd terms) has relation with the electrostatic energy:

$$\Delta P - \lambda\ \kappa(s) - P_{0n} E_{0n}(\mathbf{r}(s)) - P_{0L} E_{0L}(\mathbf{r}(s)) = 0, \qquad (1)$$

where ΔP is pressure difference between inner and outer domain, λ is line tension, $E_{0n}(\mathbf{r}(s))$ and $\mathbf{E}_{0L}(\mathbf{r}(s))$ (P_{0n} and , $\mathbf{P}_{0L}(\mathbf{r}(s))$) are the normal and in-plane components of electric field (spontaneous polarization), respectively, at the domain boundary ($\mathbf{r}(s)$) with curvature $\kappa(s)$. To solve Eq.(1) is a hard task, but employing the Taylor's expansion approach we could obtain the solution in simpler cases, such as circle and m-sided quasipolygon. MDC experiment with the BAM observation for phospholipid monolayers on the water surface have evidently showed the generation of MDC along with such predicted domain shapes by monolayer compression. As mentioned earlier, using MDC measurement coupled with BAM observation method could probe surface polarization phenomena originating from dipoles and its related surface phenomena.

3. Probing of carrier motion in organic materials as dielectric polarization phenomena

Since the discovery of conducting materials, organic material electronics has commanded increasing attention, and development of techniques that can probe carrier motion is highly motivated by scientists and electronics engineers. As mentioned in section I, it is very instructive once to trace back to the Faraday's idea. Using an EFISHG technique, we can directly probe carrier motion in organic materials [4]. The concept is to probe the propagation of polarization induced by moving carriers in organic materials. Nonlinear polarization P is induced by laser irradiation, owing to the quantum coupling between electromagnetic fields and electrons in organic materials. As electron clouds of molecules would be distorted by D.C. field, this kind of nonlinear polarization P can be induced in centro-symmetric molecular system, such as pentacene and phthalocyanine. That is, induced polarization is given as

$$P(2\omega) = \chi^{(3)} E(0) E_i(\omega) E_j(\omega), \qquad (2)$$

where ω and 0 represents the angular wave frequency of incident electromagnetic wave, E_i and E_j represent the electric field of light, and $E(0)$ is the static electric field induced from carriers. The induced polarization $P(2\omega)$ is a source of second harmonic signal, and the enhanced SH is in proportion to $P(2\omega)$ as

$$I(2\omega) \propto \left|\chi^{(3)} E(0) E_i(\omega) E_j(\omega)\right|^2 \propto \left|P(2\omega)\right|^2, \qquad (3)$$

where $I(2\omega)$ describes the SH intensity. From eqs.(2) and (3), it is clear that SH is capable of probing carrier transfer.

Fig. 1 shows the TRM-SHG image from the channel of top-contact pentacene FET with Au-source and drain electrodes [5]. The SHG image was obtained at various times after a pulse voltage of V_{pulse} =Vds=Vgs= -100 V were applied to the gate and drain electrodes. At t =0 ns, the laser pulse coincides with the rising edge of the voltage pulse, and SHG signals were found near the edge of source electrode, indicating that carrier injection just started, and a very high Laplace field was formed only around source electrode. Interestingly, as clearly shown in the image, the emission band of SHG signal gradually moved in the channel from source to drain electrode with elapsed times of 100, 200 and 400 nsec. Propagation of the emission band from source electrode, not from drain, evidently shows the hole-injection followed by hole-propagation from Au-source electrode. That is, pentacen FET shows a p-type behaviour.

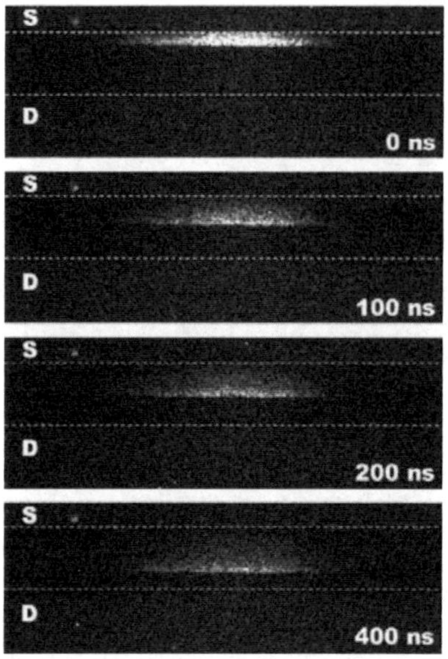

Figure 1. Experimental configurations and results of the TRM-SHG measurement.

As mentioned above, it is clear that we can probe a long-range carrier motion in organic materials by monitoring the propagation of the electric field diverging from the moving carrier. In our recent study, we could show that the peak and shoulder of SHG signals migrate along the channel in a way of square-root of time. Further the technique can apply to probing carrier motion in Organic solar cells, Organic Electroluminescence devices and so forth.

[1] Iwamoto M Majima Y Naruse H Noguchi T and Fuwa H 1991 *Nature* **353** 645
[2] Iwamoto M and Wu C X 2001 *The physical properties of organic monolayers* (Singapore: World Scientific, Singapore)
[3] Iwamoto M Manaka T Yamamoto T Lim E 2008 *Thin Solid Films* **517** 1312
[4] Iwamoto M Manaka T, Weis M, Taguchi D 2010 *J. Vac. Sci. Technol.* **B 28** C5F12
[5] Manaka T, Lim E, Tamura R and Iwamoto M 2007 *Nat. Photonics* **1** 581
[6] Iwamoto M and Ou-Yang Z C 2004 *Phys. Rev. Letts.* **93** 206101

Journal of Physics: Conference Series **358** (2012) 012011 doi:10.1088/1742-6596/358/1/012011

Fabrication of organic conductive wires and molecular break junction

Masatoshi Sakai[1], , Hiroshi Yamauchi[1], Masakazu Nakamura[2] and Kazuhiro Kudo[1]

[1]Department of Electrical and Electronic Engineering, Chiba University, 1-33 Yayoi- cho, Inage-ku, Chiba, 263-8522, Japan

[2] Graduate School of Materials Science, Nara Institute of Science and Technology, 8916-5 Takayama, Ikoma, Nara 630-0192, Japan

E-mail: sakai@faculty.chiba-u.jp

Abstract. Break junction have been proposed as one of methods for making a molecular-scale nano-gaps between the two metallic electrodes. Here we had made organic metallic wire bridges being consist of tetrathiafulvalene-tetracyanoquinodimethane (TTF-TCNQ) organic charge transfer complex, and then gradually disconnected the bridged molecular wires by a Joule heating to make a molecular-size gap or single-molecular junction. The observed conductance of the initially high conductive wires decreased by a direct current (DC) Joule heating. The conductance exhibited several different behaviors with the stage of the thermal desorption of molecules, and finally disconnected. Significant increase and decrease of the conductance were observed just before the disconnection. After the DC treatment and before the disconnection of the wires, we switched to alternation current (AC) Joule heating to make a single-molecular junction which would work as a single-molecular transistor. During the AC Joule heating, many discrete steps were found in the time evolution of the conductance. The steps were approximately 1 pS or it's multiple, and the duration of the step were nearly the same (approx. 300 - 400 s) in the final stage of the AC Joule heating. We conclude that the step height corresponds to the single-molecular conductance and the duration reflects the probability of single TCNQ molecular desorption.

1. Introduction

Many organic and inorganic nanostructures such as nano-dots, nano-wires, and nano-rods etc. have been constructed by bottom-up or self-organization methods. However, it is always difficult to make a wiring in the nanostructures. In addition, an interface double layer and depletion layer are generally formed at an interface of the semiconductor and metallic wires. Although these phenomena cause a contact resistance for the macroscopic devices, they crucially affect for the single-molecular devices because the electronic state of single molecule is fully modulated by the metallic wire contact.

We had proposed novel wire-type organic nano-transistors using a self-aligned formation techniques induced by an external electric field [1-5]. We had developed spontaneous and self-aligned formation techniques of organic nano-transistors using well-known organic conductor TTF-TCNQ [6-16]. Oriented growth and succeeding self-aligned connection were induced by an external electric field

[1] To whom any correspondence should be addressed.

during a co-evaporation of the TTF and TCNQ [4-6]. Due to the lack of TTF at the growth tip, nano TCNQ region was spontaneously formed at the junction. In this way, nano-TCNQ-transistors wired by organic conductive wires were formed[17-19]. These nano-transistor exhibited n-channel field effect transistor (FET) characteristics, which reflected the acceptor nature of TCNQ. The channel length of the spontaneously formed channel was 400 - 700 nm, which was estimated with the AFM potentiometry [20]. These nano-devices were spontaneously formed and free from carrier injection barrier in principle [17].

In this paper, we present the field-induced crystal growth of TTF-TCNQ by making use of static electric field to make connected TTF-TCNQ wires. Then we tried to disconnect these connected wires by DC Joule heating. DC Joule heating was carried out under the constant voltage operation because the constant voltage operation has a self-ceasing effect on the Joule heating. Since the self-ceasing effect was insufficient to make a single molecular junction, we applied AC Joule heating to achieve a single-molecular desorption at around several-molecular junction formed by the DC pretreatment.

2. Experimental details

Field induced selective growth of TTF-TCNQ was carried out using an in-situ observation chamber specially designed for this experiment. The in-situ observation chamber has a large quartz window at the top of the body. The growth of TTF-TCNQ is observed though the window by a long working-distance optical microscope. In this method, one can observe the growth process in real time and without any additional image processing. The shape of designed Cr/Pt electrode is a parallel electrode with several triangular tips. The applied voltage during the field induced growth was in the range of 20 – 60 V. Under a quasi-thermal equilibrium condition, TTF-TCNQ conductive wires are expected to selectively grow from these tips of the electrode and grow along the electric lines of force. With increasing the crystal length, the electric field between the growth tip of two organic conductor increases. Therefore, these organic conductors finally make a connection at each growth tips.

In the DC break junction, constant DC 1 – 3 V was applied to the connected TTF-TCNQ wires according to the sample condition. If the applied voltage is too high, the connected TTF-TCNQ wires immediately burn out. Since there is a high resistive point at the connecting point of two TTF-TCNQ wires, the proper applied voltage is fixed by the electrical resistance of the sample i.e. the connecting point. The constant voltage DC mode is useful to our purpose. Because the electrical current through the connected wires decreases with increasing sample resistance, the Joule heat automatically decreases; this method has an effect to prevent avalanche blow out of the wires in the final stage of the break junction. In addition, AC break junction is conducted to make it possible more critical Joule heating by avoiding the continuous heating. In the AC break junction, applied AC amplitude was selected in the rage of 1 -10 $V_{p\text{-}p}$, and the frequency was selected to be 7020 Hz. 7020 Hz is near the limit of a time resolution of the lock-in amplifier to measure an electrical conductance.

3. Results and Discussion

Figure 1 is an optical micrograph of grown TTF-TCNQ crystals and wire bridges around the electrode tip. The voltage applied during the growth was 40 V at the left side electrode. Tiny TTF-TCNQ crystals grew from the electrode and some wires made a connection. By an analysis of the time variation of the captured image, the preferential growth from anode and high electric field area is clearly revealed. Incubation time was the shortest and crystal growth rate was the highest in the anode and high electric field area. On the other hand in the cathode side and high electric field area, the incubation time was the longest and crystal growth rate was the lowest (to be published elsewhere). Therefore, a speculated picture of this selective growth was explained by a molecular drift motion of ionized TCNQ along the electric field. Distribution of negatively ionized TCNQ molecules is biased toward the anode by the electric field.

Next, we tried DC molecular break junction to develop a molecular-level nano-gap electrode or nano-transistor. If an electric current through the connected TTF-TCNQ wire is applied, generated Joule heat is concentrated at the high resistant connecting point. The connecting points are originally high resistant because of the lack of TTF concentration and the low cross section compared to the trunk of the wire. Since TTF has higher vapor pressure compared to TCNQ, TTF molecules at around the junction firstly vaporize, and as a result, the resistance at the junction increases. The increase of resistance causes more concentration of the Joule heat at the junction.

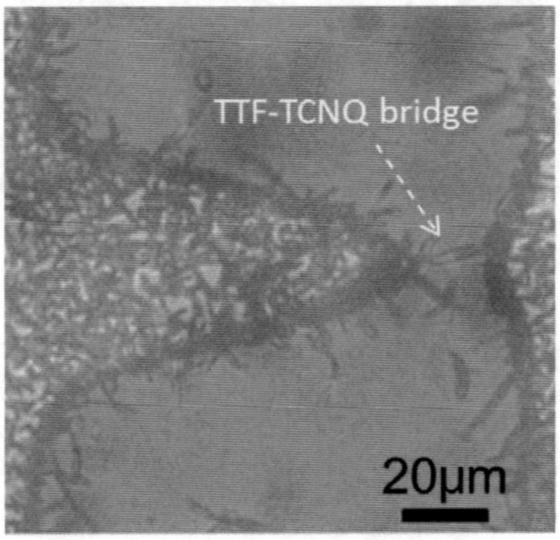

Figure 1. Optical micrograph of grown TTF-TCNQ crystals and bridges around the electrode tip.

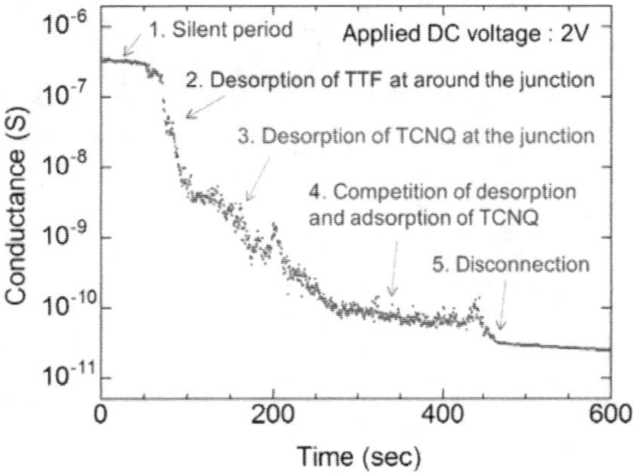

Figure 2. Time variation of the electrical conductance of TTF-TCNQ wires under the DC Joule heating.

Figure 2 shows the time evolution of the total electrical conductance of the sample under the application of DC 2 V. Characteristic decrease of the electrical conductance was observed. In the initial stage of Joule heating, the conductance slowly decreases. This period is assigned as an initial TTF desorption process, which have an effect to fix the high resistance point. Then after 70 s, steep decrease of the conductance began. This steep decrease of conductance corresponds to the main and successive TTF desorption. After 70 s, characteristic slower decay of the conductance and the significant increase and decrease which was larger than the base noise level were observed until 270 s. This stage corresponds to the TCNQ molecule desorption at the junction because the concentration of the generated Joule heat was accelerated by the vaporization of the TTF. Then, after 270 s, the conductance exhibited the very slow decay and significant increase and decrease exceeding the base noise level. After a large increase of conductance at 440 s, the TTF-TCNQ wires finally disconnected at 460 s. The nearly constant conductance with significant rises indicates that the competition of desorption and adsorption of TCNQ occurred at the heated junctions. We have found no discrete step on the conductance change even after statistical analysis due to the lack of the time resolution and the controllability of the generated Joule heat.

As an advanced trial, AC current application was conducted after the 3rd stage of the DC treatment. The advantage of the continuous and constant AC pulse train application is to prevent an avalanche burning of the junction. Figure 3 shows an example of the time evolution of the total conductance during AC Joule heating. Amplitude of the applied sine wave voltage on this sample was set to 10 $V_{p\text{-}p}$. Since continuous AC voltage generates well-controlled heat pulse train of V^2/R Joule/pulse, AC Joule heating is better than DC method to achieve a single molecular disconnection. If the generated Joule heat is comparable to or slightly lower than the thermal energy required for the single event of the TCNQ adsorption, no single event occurs within the single heat pulse. Under this condition, the number of the heat pulse reflects the probability of the single molecular desorption under the supplied thermal energy. In figure 3, discrete steps of total conductance are observed. The difference of the conductance at each step is approximately 1 pS or it's multiple as indicated in figure 3. Moreover in the final stage, the duration of one step (approximately 300 - 400 s) is almost the same in each step.

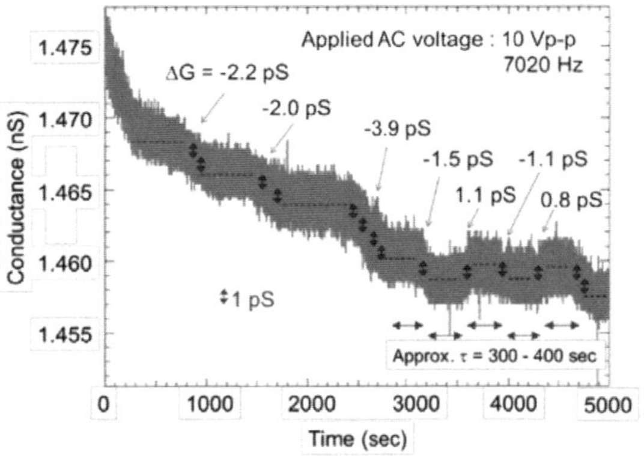

Figure 3. Discrete decrease and increase of the conductance under the application of the AC Joule heating.

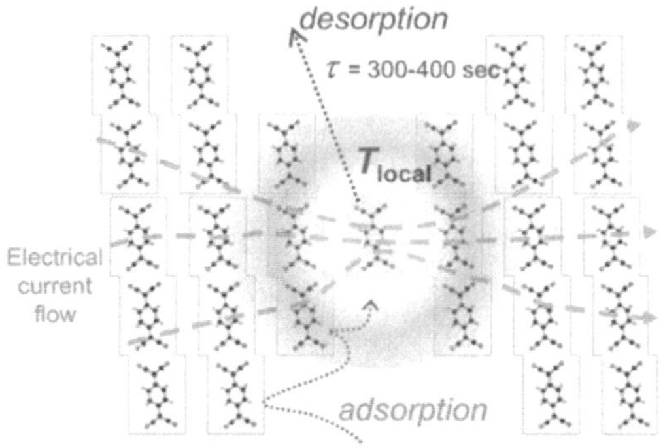

Figure 4. Schematic illustration of the single molecular bridge. An electronic current flowing through the connected wires is concentrated at the molecular bridge.

The observed discrete increase and decrease of the conductance correspond to the single molecule desorption at molecular junctions. Due to our experimental results, single molecular desorption occurs in every 300 – 400 s i.e. $2.1 - 2.8 \times 10^6$ pulses. In the final stage of the Joule heating, an electrical current is concentrated at the single molecular bridge as illustrated in figure 4. The heat pulse generated by the single current pulse defines the effective local temperature (T_{local}) at the molecular bridge. The single event of molecular desorption stochastically occurs every $2.1 - 2.8 \times 10^6$ pulses. The assumed T_{local} is now under estimation by the molecular dynamics calculation. In addition, it is supposed that there exist many parallel molecular bridges because the conductance oscillation observed in this experiment will cease if the last molecular bridge burn out. In our rough estimation, 20 - 50 molecular junctions will parallel exist.

4. Summary

We had made organic metallic wire bridges of TTF-TCNQ organic charge transfer complex, and then gradually disconnected them by DC and AC Joule heating to make single-molecular junctions. The observed conductance under DC Joule heating exhibited several different behavior with the stage of the thermal desorption of molecules, and finally disconnected. After the DC treatment and before the disconnection of the wires, we switched to AC Joule heating to make a single-molecular junction. During the AC Joule heating, many discrete steps were found in the time evolution of the conductance. These steps were approximately 1 pS or it's multiple, and the duration of the step were approximately 300 - 400 s in the final stage of the AC Joule heating. We have concluded that the step height corresponds to the single-molecular conductance and the duration reflects the probability of single molecular desorption.

Acknowledgments

This work was supported by Grant-in aid for Scientific Research and Chiba University Global COE Program from Ministry of Education, Culture, Sports, Science and Technology (MEXT), Japan. Part of this study was conducted by the support from the "Toyo University nanotechnology Network", which comprises the "Nanotechnology Network Japan" by MEXT.

References

[1] Sakai M, Iizuka M, Nakamura M and Kudo K 2004 Thin Film Transistors with Oriented Copper Phthalocyanine Crystals Fabricated by Physical Vapor Deposition under DC Electric Field *Jpn. J. Appl. Phys.* **43** 2362

[2] Sakai M, Iizuka M, Nakamura M and Kudo K 2011 Visible effects of static electric field on physical vapor growth of lead phthalocyanine crystals *J. Appl. Phys.* **109** 054309
[3] Sakai M, Iizuka M, Nakamura M and Kudo K 2005 Self-organized growth of tetrathiafulvalene-tetracyanoquinodimethane molecular wires using the co-evaporation method under a static electric field *J. Appl. Phys.* **97** 053509
[4] Sakai M, Iizuka M, Nakamura M and Kudo K 2003 Fabrication and Electrical Characterization of Tetrathiafulvalene-tetracyanoquinodimethane Molecular Wires *Jpn. J. Appl. Phys.* **42** 2488
[5] Kato N A, Fujimura M, Kuniyoshi S, Kudo K, Hara M and Tanaka K 1998 Control of Molecular Orientation of TTF-TCNQ Co-evaporated Films by an Applied Electric Field *Appl. Surf. Sci.* **130-132** 658
[6] Phillips T E, Kistenmacher T J, Ferraris J P and Cowan D O 1973 Crystal structure of the radical-cation radical-anion salt from 2,2′-bi-1,3-dithiole and 7,7,8,8-tetracyanoquinodimethane *J. Chem. Soc. Chem. Commun.* 471
[7] Kitstenmacher T J, Phillips T E and Cowan D O 1974 The crystal structure of the 1:1 radical cation-radical anion salt of 2,2'-bis-l,3-dithiole (TTF) and 7,7,8,8-tetracyanoquinodimethane (TCNQ) *Acta Cryst.* **B30** 763
[8] Berlinsky A J, Carolan J F and Weiler L 1974 Band structure parameters for solid TTF-TCNQ *Solid State Commun.* **15** 795
[9] Salahub D R, Messmer R P and Herman F 1976 Study of tetrathiafulvalene (TTF) dimers and trimers in eclipsed and slipped geometries *Phys. Rev.* **B13** 4252
[10] Cohen M J, Coleman L B, Garito A F and Heeger A J 1974 Electrical conductivity of tetrathiofulvalinium tetracyanoquinodimethan (TTF) (TCNQ) *Phys. Rev.* **B10** 1298
[11] Etemad S 1976 *Phys. Rev.* **B13** 2254
[12] Ishiguro T, Kagoshima S and Anzai H 1976 Systematic study of the transitions in tetrathiafulvalene-tetracyanoquinodimethane (TTF-TCNQ) and its selenium analogs *J. Phys. Soc. Jpn.* **41** 351
[13] Cohen M J and Heeger A J 1977 Nonlinear transport in pinned one-dimensional charge-density wave systems *Phys. Rev.* **B16** 688
[14] Cooper J R 1979 Comments on the metallic conductivity of tetrathiafulvalenium-tetracyanoquinodimethanide (TTF-TCNQ) *Phys. Rev.* **B19** 2404
[15] Bright A A, Garito A F and Heeger A J 1973 Optical properties of (TTF) (TCNQ) in the visible and infrared *Solid State Commun.* **13** 943
[16] Tanner D B, Jacobsen C S, Garito A F and Heeger A J 1976 Infrared studies of the energy gap in tetrathiofulvalene-tetracyanoquinodimethane (TTF-TCNQ) *Phys. Rev.* **B13** 3381
[17] Sakai M, Nakamura M and Kudo K 2007 Organic nanochannel field-effect transistor with organic conductive wires *Appl. Phys. Lett.* **90**, 062101
[18] Sakai M, Iizuka M, Nakamura M and Kudo K 2005 Organic nano-transistor fabricated by co-evaporation method under alternating electric field *Synth. Met.* **153** 293
[19] Sakai M, Miyata H, Itami K, Nakamura M and Kudo K 2008 Spontaneous Activation Process for Self-aligned Organic Nanochannel Transistors *Appl. Phys. Express* **1** 081802
[20] Nakamura M, Fukuyo M, Wakata E, Iizuka M and Kudo K, Tanaka K 2003 Development of AFM Potentiometry for Potential Mapping of Organic Conductors *Synth. Met.* **137** 887

Journal of Physics: Conference Series **358** (2012) 012012 doi:10.1088/1742-6596/358/1/012012

Solution-phase laser processing of π-conjugated polymers: Switching between different molecular states

K Takada and A Tomioka[1]
Graduate School of Engineering, Osaka Electro-Communication University
18-8 Hatucho, Neyagawa, Osaka 572-8530, Japan

E-mail: tomioka@isc.osakac.ac.jp

Abstract. Liquid-phase laser processing, where the laser-irradiated target material is immersed in water for cooling, has been reported as a promising processing technique for thermally fragile organic materials. Although nanometer-sized particles have been reported to be obtained with the liquid-phase laser processing, the physical property did not change because quantum-mechanical size effect does not exhibit itself in the zero-radius Frenkel excitons. In the present study, we step further to use solution droplets as a target material, where organic molecules are molecularly dispersed in organic solvent and, therefore, expected to easily alter the conformation and the energy state upon laser irradiation. Small volume organic solvent is quickly evaporated upon laser irradiation, letting the bare organic molecule placed in water and rapidly cooled. To prevent the chemical decomposition of the target π-conjugated molecule, the specimen was resonantly irradiated by a ns-pulse green laser, not by a conventional UV laser.

When the solid state spin-coat film made from MEH-PPV chloroform solution was used as a irradiation target immersed in water, resulting MEH-PPV particles showed similar photoluminescence (PL) like the PL of the spin-coat film and PL of the chloroform solution, including the $0\rightarrow1$, $0\rightarrow2$ vibrational transitions: this indicates that the energy levels were not modified from the spin-coat film. In comparison, when tiny droplets of MEH-PPV chloroform solution (orange color) were suspended in water, laser irradiation gave rise to yellow MEH-PPV particles which showed 550 nm and 530 nm PL (type B), blue-shifted from the spin-coat film PL 580 nm (type A), suggesting a successful phase transition of MEH-PPV polymer to type B. Further solution-phase laser processing left the type B state unchanged. The irreversible phase transition from type A to type B suggests that the type B ground state has lower energy than type A, which is consistent with the blue-shifted PL of type B, provided that the excited state energy is similar between the two states. Thermal annealing up to 200°C of type A state did not give rise to type B state, which indicates that the activation potential between the two states is higher than the thermal energy at 200°C, and that only the proposed solution-phase laser processing enables the system to cross over this potential.

1. Introduction

Laser ablation [1] of solid materials is convenient and efficient technology to fabricate a material with designed chemical composition. Since the technology utilizes a high-photon-energy UV laser to

[1] To whom any correspondence should be addressed.

decompose the original material into atom clusters, it is inapplicable to fragile organic material whose properties are governed by molecular structure. Liquid-phase laser processing (LLP) was proposed recently [2], where the laser-irradiated solid material is immersed in water in prospect of cooling effect of water in order to avoid degradation of organic materials. Although this technique brought about nanometer-sized particles, the physical and chemical properties of the particles were the same as the original material because no quantum size effect was expected in the case of Frenkel bound excitons with zero Bohr radii. A novel solution-phase laser processing (SLP) is proposed in the present research, where tiny droplets of molecularly dispersed material solution are adopted as the laser-irradiation target and homogeneously suspended in insoluble water. Molecularly-dispersed material is expected to have more possibility to attain a phase transition, free from the binding energy between molecules.

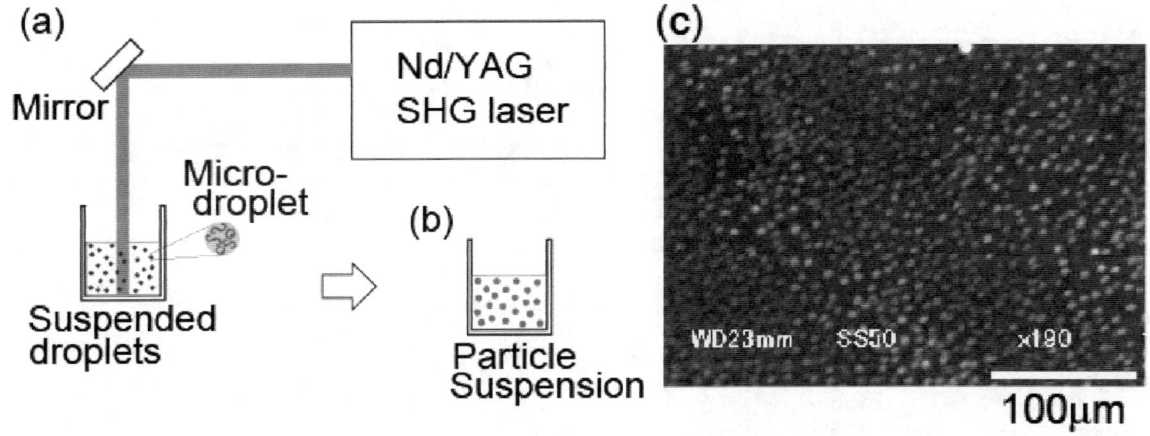

Figure 1. (a) Schematic drawing of SLP processing where a low-photon-energy green pulse laser is irradiated against the target, i.e. polymer solution microdroplets suspended in insoluble water. (b) Resulting polymer particles suspended in water whose SEM image is shown in (c) with a 100 µm scale bar.

Target material is a π-conjugated dye polymer, poly[2-methoxy-5-(20-ethyl-hexyloxy)-pphenylene-vinylene] (MEH-PPV), which has strong optical absorption at visible wavelength. Upon resonant irradiation of a green pulse laser, molecular motion is expected to be invoked via the strong electron-phonon coupling of the extended π-conjugated system [3 - 5], leading to a plausible phase transition of the polymer conformation [6]. Low-photon-energy visible pulse laser would be effective to invoke this transition, without any chemical damage of the material.

2. Materials and methods

An aliquot of 200µl MEH-PPV solution in chloroform ($CHCl_3$), dichloromethane (CH_2Cl_2), chlorobenzene (C_6H_5Cl), or toluene ($C_6H_5CH_3$), was put in 2 ml distilled water and ultrasonicated for 15 min. After microdroplets of MEH-PPV solution were uniformly suspended in water (turbid appearance due to light scattering), collimated Q-switched Nd/YAG SHG laser (λ=532nm, 2 mJ 5ns pulse, 10Hz, Minilite II, Continuum) was irradiated for typically 30min as a SLP processing (figure 1(a)). The resulting homogeneous suspension (figure 1(b)) was dried on a silicon wafer, vacuum evaporated with gold, and observed by scanning electron microscope (SEM, JSM-6510SLV, JEOL) at lower accelerating voltage of 20kV in order to avoid charging up with incident electrons. The photoluminescence (PL) was observed under a microscope and its spectrum was recorded by a spectrometer attached with a charge coupled device camera with on chip multiplication capability (SP-2358 + PhotonMAX 97EMB, Roper Scientific). The thermal annealing experiment up to 200°C was carried out with the specimen held in a sealed chamber filled with dried N_2 gas.

3. Results and discussions

3.1. Conventional liquid-phase laser processing

For comparison, conventional LLP processing was tried first, with a solid phase irradiation target: MEH-PPV spin-coat film made from chloroform solution. The apparent color of the resultant suspension in water did not change before and after the LLP processing. The photon energies of the PL main peak and its vibronic subpeaks were similar with those of MEH-PPV chloroform solution (figure 2(a)), which indicates that the electronic state of MEH-PPV was unchanged. We name this PL as type A, in this paper.

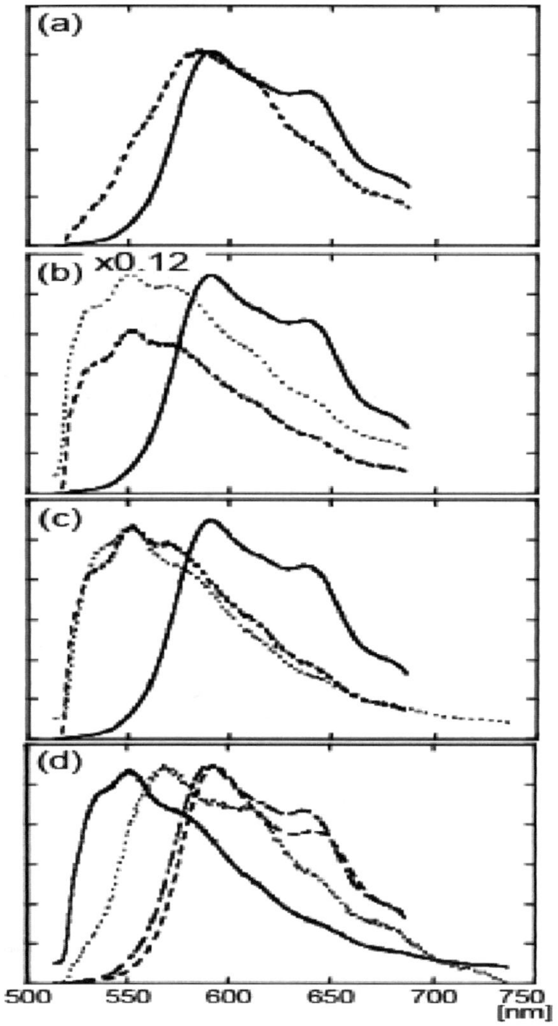

Figure 2. Comparison of PL spectra. (a) spin-coat film (solid line) and LLP processed spin-coat film (dashed line) made from chloroform solution. (b) spin-coat film (solid line) and SLP processed particle before (dashed line) and after the annealing (dotted curve, diminished by 0.12 times), made from chloroform solution. (c) spin-coat film (solid line) and SLP processed and annealed particle (dashed line) made from chloroform solution. Completely overlapped two dotted curves are a spin-coat film, and SLP processed / annealed particle, made from dichloromethane solution. (d) Solution PL dissolved in dichloromethane (solid line), chlorobenzene (dotted line), chloroform (dashed-dotted line), and toluene (dashed line).

3.2. Solution-phase laser processing

In SLP processing, microdroplets of MEH-PPV chloroform solution were uniformly suspended in water and used as a laser irradiation target. Since tiny volume of volatile solvent chloroform are expected to rapidly evaporate upon laser irradiation, insoluble MEH-PPV polymer will be cooled and form tiny particles in water. After SLP processing, the apparent color of the suspension was changed from orange to yellow for chloroform, toluene, and chlorobenzene solution, while the color was already yellow in the case of dichloromethane solution.

The resulting homogeneous suspension was found to be composed of 5μm MEH-PPV particles by the SEM observation (figure 1(c)). The size of the particles seemed to be homogeneous and no aggregation was observed, which is a promising feature for application.

The PL was blue-shifted from that of spin-coat film (figure 2(b)): two new peaks emerged at 550 nm and 530 nm. We call this PL as type B, in the following. The PL intensity was increased after thermal annealing up to 200°C, with intensity profile unchanged. When dichloromethane was used as the starting solvent, the spin-coat film and the SLP processed particles both showed the identical type B spectrum as shown in figure 2(c) dotted curve. Dichloromethane solution of MEH-PPV also showed type B PL, whereas chloroform and toluene solution showed type A PL, as shown in figure 2(d). Chlorobenzene solution showed PL with maximum in between: it may be a mixture of many states because it also seemed to show both type A and type B PL.

In all cases, SLP processing forced the specimen into type B state irreversibly, which suggests that type B state has lower (ground state) energy than type A state (figure 3). This interpretation is consistent with the fact that type B state PL was blue-shifted from that of type A state, provided that the excited state energy is similar between the two states. The activation potential between the two states should be higher than the thermal energy at 200°C, because thermal annealing up to 200°C of type A state did not give rise to type B state. The fact that dichloromethane solvent can induce this transition suggests that this transition may reflect some conformational change of MEH-PPV polymer itself, not a change of the packing of polymer molecules in the particles. This is not a solvatochromism, because the PL showed type B spectrum even after dichloromethane evaporated completely from the spin-coat film by thermal annealing: instead, the transition can be induced by dichloromethane solvation and is irreversible.

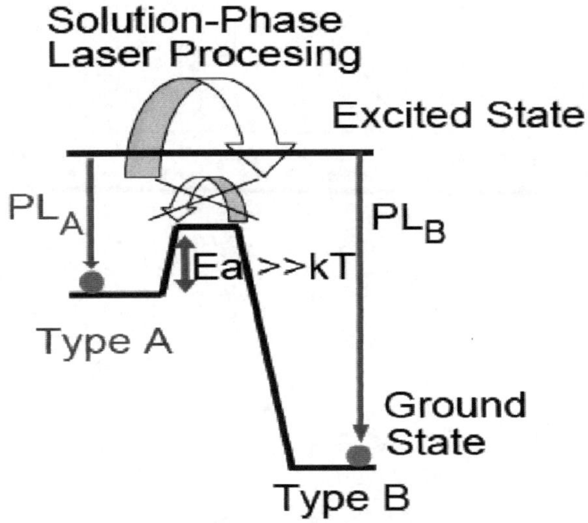

Figure 3. (a) Schematic model of the phase transition. SLP processing irreversibly converts the MEH-PPV type A state to type B state which has lower ground state energy. This stabilization is consistent with the blue-shifted PL of type B state. The activation potential E_a between the two states is higher than the thermal energy kT at 200°C.

In figure 2, every PL of type B is accompanied with type A peaks. It may mean that these spectra are mixture of type B and type A state. In fact, with shorter term SLP processing of chloroform solution, type A peaks remained as major components, larger than type B peaks, which suggests that the two states type A and type B belong to different electronic states and that they coexist in the same specimen with a ratio dependent on the degree of SLP processing.

In order to evaluate the electronic transition energy and the vibrational energy, PL spectra were replotted with the photon energy as abscissa, and deconvoluted into different vibrational transitions (figure 4). PL spectra were fitted mathematically as a superposition of Gaussian peaks at photon energy E_i,with linewidth ΔE_i:

$$I_{PL}(E) = \sum_i I_i \exp\left[-\frac{(E-E_i)^2}{\Delta E_i^2}\right] \tag{1}$$

Since the Lorentzian function form was not fitted well with the PL spectrum, Gaussian function was adopted successfully. This fact suggests that the present inhomogeneous linewidth may be governed by static and dynamic, i.e. transient polymer deformation due to vibrational mode, disorders, similar to the case of energy transitions lower than the band gap of pseudo-isocyanine dye J-aggregate [7, 8].

Table 1. Deconvoluted PL peak energies corresponding to different vibrational transitions. Specimens are sorted by the original solvent in which MEH-PPV was dissolved prior to the processing: (a) spin-coating, (b) SLP processing.

(a)

solvent	Δν	λ[nm]	E[eV]	ΔE[eV]
chloroform	0→0	590	2.10	
	0→1	642	1.93	-0.17
toluene	0→0	593	2.09	
	0→1	646	1.92	-0.17
dichloro-		528	2.35	0.23
methane		549	2.26	0.14
	0→0	593	2.09	
	0→1	626	1.98	-0.11
	0→2	653	1.90	-0.19

(b)

solvent	Δν	λ[nm]	E[eV]	ΔE[eV]
chloroform		528	2.35	0.21
		549	2.26	0.12
	0→0	579	2.14	
	0→1	614	2.02	-0.12
	0→2	639	1.94	-0.20
toluene		530	2.34	0.20
		556	2.23	0.09
	0→0	579	2.14	
	0→1	611	2.03	-0.11
	0→2	653	1.90	-0.24
dichloro-		530	2.34	0.26
methane		551	2.25	0.17
	0→0	596	2.08	
	0→1	626	1.98	-0.10
	0→2	653	1.90	-0.18

The spin-coat film made from MEH-PPV chloroform solution and that from toluene solution both showed type A emission (figure 4(a) and (c), respectively) which are well reproduced in the photon energy range above 1.85eV by a superposition of two Gaussian peaks, corresponding to 0→0 (E_{A0}) and 0→1 (E_{A1}) vibrational transitions [9]. Discrepancy around 1.8eV suggests that it is due to a small contribution of additional 0→2 vibrational transition. From this decomposition, the electronic transition energy was deduced to be 1.93eV and the vibrational energy was 0.17eV, consistently between two specimens (table 1(a)). In the case of the spin-coat film made from MEH-PPV dichloromethane solution (figure 4(b)), type B emissions at E_{B1}=2.26eV and E_{B2}=2.35eV showed up, accompanying type A emissions, E_{A0}, E_{A1}, E_{A2} as minor components. The spacing E_{B1} to E_{B2} is 0.09eV, smaller than the vibrational energy 0.17eV, suggesting that type B emissions E_{B1} and E_{B2} have origins different from type A emissions.

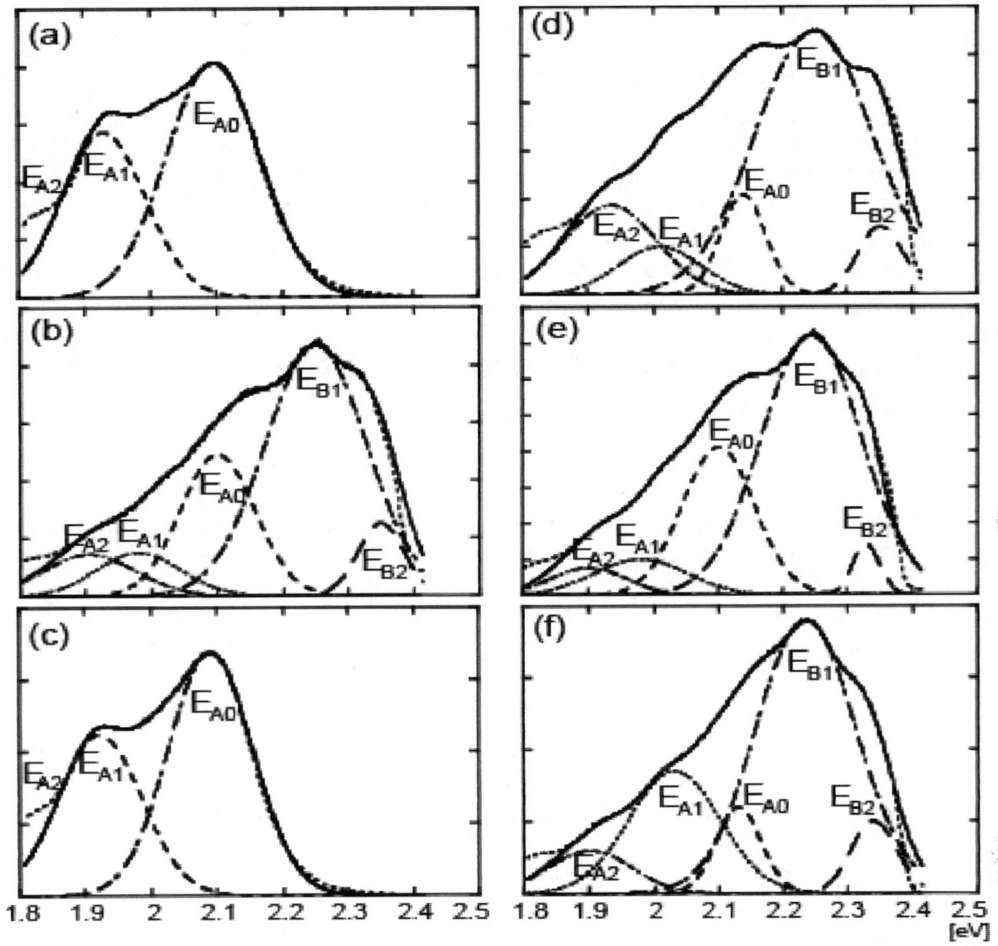

Figure 4. Deconvolution of PL spectra into different vibrational transitions. The abscissa is expressed in photon energy so as to show clearly the equispaced energies among different vibrational transitions (E_{A0}, E_{A1} and E_{A2}) Decomposed peaks are drawn by dashed or dashed-dotted lines. The mathematical superposition of these decomposed peaks is expressed as thick solid curves which overlap the original PL spectra shown by dotted lines. (a)-(c): spin-coat films, (d)-(f): SLP processed particles. (a), (d): made from chloroform solution, (b), (e): from dichloromethane solution, (c), (f): from toluene solution

After SLP processing, the same PL spectrum resulted for all the particles made from chloroform, dichloromethane, and toluene solution (figures 4(d), (e), and (f), respectively). Majority of the PL belongs to type B emissions E_{B1} and E_{B2}, although minor components E_{A0}, E_{A1}, and E_{A2} remained, with the observed PL spectrum well-reproduced by the superposition of E_{Ai} and E_{Bi} five components.

Judging from figure 4, relative intensities E_{A1} / E_{A0}, E_{A2} / E_{A0} of the deconvoluted components are different among different solvents, whereas they are similar between the spin-coat film and the SLP particles made from the same dichloromethane solvent. This observation suggests that the electron- phonon coupling is sensitive to the solvent and that it memorizes the change even after the solvent is evaporated: one of the candidates of this "memory" effect is the conformational change of the polymer molecule.

4. Conclusion

A novel visible laser processing technique, where π-conjugated polymer solution droplets suspended in insoluble water were photoexcited resonantly by a low-photon-energy green pulse laser, was demonstrated to yield polymer particles with altered photoluminescence, which is indicative of phase transition of the polymer conformation. The final state was the same among laser processing of polymer in different original solvents, chloroform, toluene, dichloromethane. The electronic ground state of the resultant polymer particles is expected to have lower energy than the original state, which is consistent with the blue-shift of the photoluminescence after this laser processing, and also consistent with this irreversible phase transition upon the laser processing. Thermal annealing up to 200 °C did not give rise to this phase transition, which indicates that the activation potential between the two states is higher than the thermal energy at 200 °C, and that only the proposed solution-phase laser processing enables the system to cross over this potential and to transit to the same state, irrespective of the molecular environment defined by surrounding solvents.

Reference

[1] Lucatorto T and Graef M D 1997 *Laser Ablation and Desorption* vol 30 (Experimental Methods in the Physical Sciences), ed R F Haglund and J C Miller (Academic Press)
[2] Bauerle D 2011 Liquid-Phase Deposition, Electroplating *Laser Processing and Chemistry* (Springer-Verlag) Chap 11
[3] Liess M, Jeglinski S, Vardeny Z V, Ozaki M, Yoshino K, Ding Y and Barton T 1997 *Phys. Rev.* B **56** 15712
[4] Abe S 1989 *J. Phys. Soc. Jpn.* **58** 62
[5] Ogawa T and Takagahara T 1991 *Phys. Rev.* B **44** 8138
[6] Zhang J, Wang Z and Kobayashi T 2008 *Phy. Rev.* B **77** 153202
[7] Nabetani A, Tamaru H, Tomioka A and Miyano K 1995 *J. Chem. Phys.* **102** 5109
[8] Tomioka A and Miyano K 1996 *Phys. Rev.* B, **54** 2963
[9] Pickler K, Halliday D A, Bradley D D C, Burn P L, Friend R H, and Holmes A B 1993 *J. Phys.: Condens. Matter* **5** 7155

Acknowledgements

We express our gratitude to Professor Nobuhito Ohno, Osaka Electro-Communication University, for usage of a sealed annealing chamber for microscope observation. We would like to show our gratitude to Associate Professor Yoshiaki Yuguchi, Osaka Electro-Communication University, for usage of a liquid-cell spectrophotometer. This work was supported partly by Grant-in-Aid for Scientific Research (C), and a grant for graduate student in Osaka Electro-Communication University.

Effects of thermal annealing on structure, morphology and optoelectronic properties of TiOPc ultrathin films

R Ye[1, †], D Xiong[2], M Yanagida[2], K Ohta[1], T Abe[3] and M Baba[1]
[1]Faculty of Engineering, Iwate University, Morioka 0208551, Japan
[2]Graduate School of Engineering, Iwate University, Morioka 0208551, Japan
[3]Iwate Industrial Research Institute, Morioka 0200852, Japan

E-mail: ye@iwate-u.ac.jp

Abstract. This article reports the effects of thermal annealing on structure, morphology and optoelectronic properties of titanyl phthalocyanine (TiOPc) ultrathin films grown at room temperature. When the thermal annealing temperature (T_A) is over 160 °C, a peak at 2θ =7.5° with d_{010} = 1.194 nm was observed, which suggests that TiOPc molecules were restructured and crystalline α-form films were formed. By the detailed x-ray diffraction analysis, the optimum thermal annealing conditions are as follows: T_A is 160 °C, and the thermal annealing time is 4 h. The root-mean-square roughness and grain size are also increased due to recrystallization with increasing T_A. Similarly, these absorption spectra of TiOPc ultrathin films reveal the gradual formation of the crystalline α-phase, as evidenced by the development of a red-shifted absorption band, peaking 845 nm, and the corresponding reduction of the 720 nm amorphous phase peak when T_A was increased from 120 °C to 180 °C. Furthermore, device performance of TiOPc thin film transistors could be improved and mobility was enhanced by thermal annealing, which originated from amorphous TiOPc films transformed into crystalline α-TiOPc films and increasing of grain size. At T_A = 160 °C, the device achieved the highest performance with field-effect mobility of 6.34 x 10^{-3} $cm^2V^{-1}s^{-1}$.

1. Introduction

Metal phthalocyanines (MPcs) are organic dye materials, used in printing inks, as colorants for plastics and fibers, and their semiconducting properties are exploited for applications such as thin film transistors (TFTs), light emitting diodes, solar cells and gas sensors.[1-16] Titanyl phthalocyanine (TiOPc), one of non-planar phthalocyanines, has five polymorphs with different preparation and treatment procedures.[8-16] Amorphous or nearly amorphous TiOPc films were obtained for vacuum deposition at substrate temperature below 0 °C while β-form (phase I) or a mixture of α-form (phase II) and β-form polycrystalline films at substrate temperature above 20 °C and crystallinity increased with increasing of substrate temperature, and at substrate temperature of above 170 °C only α-form films were obtained. Moreover, amorphous TiOPc films could be transformed into crystalline α-TiOPc film through thermal annealing or vapor exposure. [13, 15] It has been known that α-TiOPc presents a particularly efficient charge photogeneration and a strong optical absorption in the red and near-IR spectral regions and a very promising organic semiconductor with high mobility due to showing the pi-stacking structure with concave pair and convex pair with significant molecular overlaps and very

† To whom any correspondence should be addressed.

short intermolecular distance.[16] However, few studies have been carried out on the correlation between structures and optoelectronic properties of TiOPc films. In this study, we report on effects of thermal annealing on structure, morphology and optoelectronic properties of TiOPc ultrathin films.

2. Experimental details

TiOPc thin films of ca. 30 nm were deposited on heavily n-doped Silicon substrates with a 300 nm thermally gown SiO_2 layer (C_i = ~10 nF/cm^2) at room temperature and under a base pressure of less than 1×10^{-3} Pa. Film thickness and growth rates were monitored by a thickness and rate monitor (CRTM-6000, ULVAC). Samples were annealed in Ar-ambient and dark conditions at temperatures ranging from 120 °C, 140 °C, 160 °C to 180 °C, respectively. The XRD analysis was performed on a diffractometer (Rint 2200V, RIGAKU Co., Ltd.) with graphite monochromatized CuK_α radiation (λ=1.54056 Å), operating in the Θ-2Θ mode. The morphology of the films was examined using AFM (Nanocute, Seiko Instruments Co., Ltd.); the cantilevers were used in the tipping mode with a length of 90 μm and a force constant of 0.12 N/m. The grain size (GS) and the root-mean-square roughness (RMS) were given by the AFM instrument for the individual scans (2 μm × 2 μm). For valuating electronic properties of top-contact TiOPc TFTs, Au source and drain electrodes of ca. 30 nm were vacuum-deposited through a shadow mask with a channel width of 5 mm and length of 70 μm. The electrical characteristics of devices were measured using a two channel voltage current source/ monitor system (R6245, Advantest) under ambient laboratory air conditions. Furthermore, TiOPc thin films of ca. 30 nm were also deposited on glass substrates for optical measurements. The absorption spectra were recorded with UV/VIS spectrophotometer (V-550, JASCO) at room temperature.

3. Results and Discussion

Figure 1(a) shows XRD patterns of TiOPc ultrathin films of ca. 30 nm before and after annealed at various temperature for 4 h. The as-deposited film was in amorphous state. When the annealing temperature (T_A) is over 160 °C, a peak at 2θ = 7.5° with d_{010} = 1.194 nm was observed, which suggests that TiOPc molecules were restructured and crystalline α-phase films were formed. However, the full width at half maximum (FWHM) of the 010 diffraction line of the films annealed at 160 °C is smaller than that of the films annealed at 180 °C. From the detailed XRD analysis of Figs. 1(a) and (b), the dependence of the ratios of integration intensity (RII) of the (010) peak on T_A and time were shown in Fig. 1(c). Thus the optimum thermal annealing conditions are as follows: T_A is 160 °C and the thermal annealing time is 4 h.

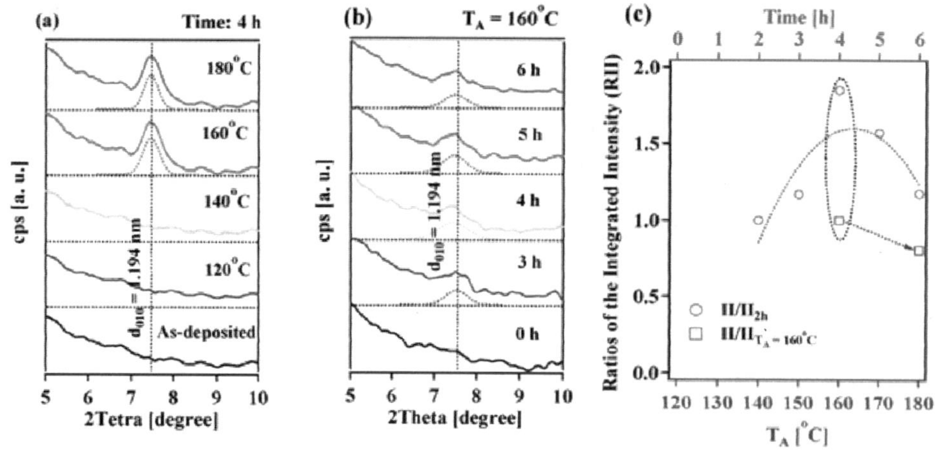

Figure 1. (a) XRD patterns of TiOPc ultrathin films before and after annealed at 120 °C, 140 °C, 160 °C and 180 °C, respectively, and for 4 h; (b) XRD patterns of TiOPc thin films annealed at 160 °C for various times; (c) The Dependence of RII of the (010) peak on T_A and time.

Figure 2 shows the 2 μm × 2 μm AFM topographic images of TiOPc thin films before and after the thermal annealing for 4 h. In Fig. 2(a), many small projections are observed, and in Fig. 2(b), some adjacent grains are joined together, which suggests that the local recrystallization has occurred due to the thermal annealing. As can be seen in Fig. 2(c), the grain size is increased and the density is decreased with increasing T_A. Furthermore, the local reevaporation could be observed in Fig. 2(d).

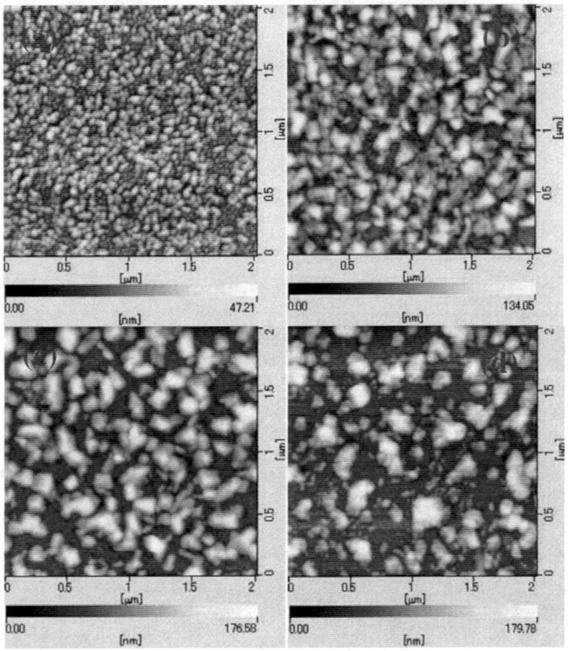

Figure 2. Topographic AFM Images of TiOPc ultrathin films before and after annealed at various temperatures for 4 h.

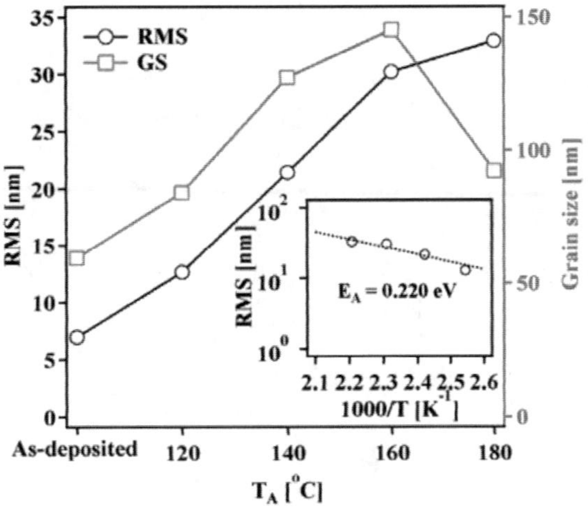

Figure 3. The RMS and GS are shown as a function of T_A. The inset shows the RMS amplitude (σ) vs. 1000/T with a linear fit yielding the activation barrier of $E_A = 0.220$ eV.

Figure 3 shows the RMS and grain size increase with increasing T_A. While T_A is over 180 °C, the grain size was decreased due to reevaporation. The RMS roughness amplitude (σ) increases closely following an Arrhenius behavior as $\sigma \propto \exp(-E_A / k_B T)$ with the activation energy E_A = 0.220 eV, and k_B is the Boltzman constant in the inset of Fig. 3. The value of E_A is smaller than that of oligomer thin films grown on the elevated substrates. This suggests that the thermal annealing caused a local aggregation of TiOPc molecules which led to an increase in the RMS, unlike pure thermal diffusion growth on elevated substrates. [17]

Figure 4 show absorption spectra of TiOPc ultrathin films prepared as-deposited and thermal annealing form 120 °C to 180 °C. The as-deposited film was amorphous and its absorption spectrum consists of the Q-band peaking at 720 nm with a well defined vibronic replica at 653 nm. When T_A was increased from 120 °C to 180 °C, these sequence absorption spectra reveal the gradual formation of the crystalline α-phase, as evidenced by the development of a red-shifted absorption band, peaking 845 nm, and the corresponding reduction of the 720 nm amorphous phase peak. The 845 nm peak, which is a spectroscopic signature of α-phase, has been attributed either to excitonic delocalization induced by the formation of J aggregates in the crystalline phase or to molecular distortion lifting the degeneracy of the fundamental electronic transition. This result is coincided with that of the XRD data.

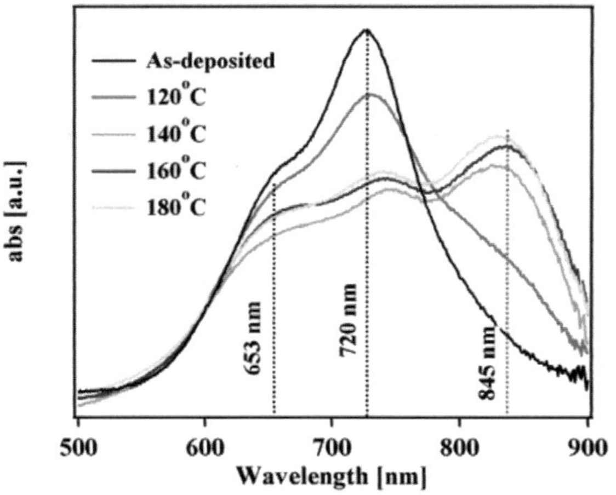

Figure 4. Absorption spectra of TiOPc thin films before and after annealed at various temperatures for 4 h.

Figure 5 shows output characteristics of TiOPc TFTs before and after annealed at 160 °C for 4 h. These devices typically work in *p*-channel accumulation mode. Compared with the as-deposited device, the saturated drain current of the annealed device at $V_D = V_G$ = -60 V was enhanced up to over 3 times. Figure 6 shows the transfer characteristics of TiOPc TFTs before and after annealed at various temperatures for 4 h. Generally, the device performance of TFTs is evaluated with field-effect mobility (μ) and threshold voltages (V_T) extracted using the saturated drain current I_D vs. V_G relation [18]:

$$I_D = \frac{W}{2L} C_i \mu (V_G - V_T)^2, \qquad (1)$$

where W, L and C_i are channel width, channel length and gate dielectric capacitance per unit area, respectively. From Fig. 6, the dependence of mobility on T_A is shown in the inset of Fig. 6. With increasing T_A, the device performance was improved. At T_A = 160 °C, the device achieved the highest performance with mobility of 6.34 x 10^{-3} $cm^2V^{-1}s^{-1}$. But the device mobility is decreased at T_A = 180 °C. The enhancement of mobility originated from the phase transition from amorphous TiOPc to α-TiOPc with an ultra closely pi-stacked structure and increasing of grain size.

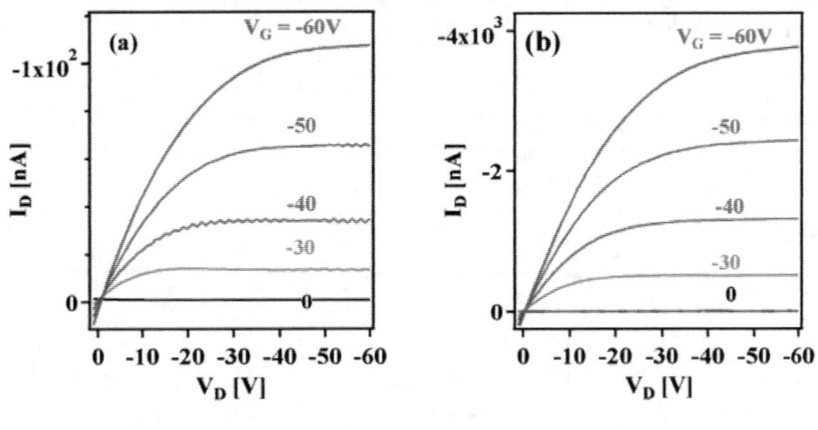

Figure 5. Output characteristics of TiOPc TFTs before (a) and after (b) annealed at 160 °C for 4 h.

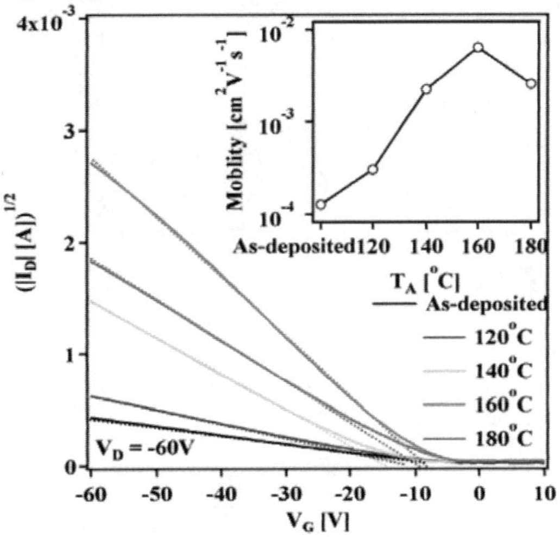

Figure 6. Transfer characteristics ($|I_D|^{1/2} - V_G$) of TiOPc TFTs before and after annealed at various temperatures for 4h. The inset shows dependence of mobility of the TiOPc TFTs on T_A.

4. Conclusion

We have investigated the effects of thermal annealing on structure, morphology and optoelectronic properties of TiOPc ultrathin films grown at room temperature. When T_A is over 160 °C, a peak at 2θ = 7.5° with d_{010} = 1.194 nm was observed, which suggests that TiOPc molecules were restructured and crystalline α-form films were formed. By the detailed XRD analysis, the optimum thermal annealing conditions are as follows: T_A is 160 °C and the thermal annealing time is 4 h. The RMS and grain size are increased due to recrystallization with increasing T_A. Similarly, these absorption spectra of TiOPc ultrathin films reveal the gradual formation of the crystalline α-phase, as evidenced by the development of a red-shifted absorption band, peaking 845 nm, and the corresponding reduction of the 720 nm amorphous phase peak when T_A was increased from 120 °C to 180 °C. Furthermore, device performance of TiOPc TFTs could be improved and mobility was enhanced by thermal annealing, which originated from amorphous TiOPc films transformed into crystalline α-TiOPc film and increasing of grain size. At T_A = 160°C, the device achieved the highest performance with field-effect mobility of 6.34 x 10^{-3} $cm^2V^{-1}s^{-1}$.

References

[1] Guillaud G, Simon J and Germain J P 1998 *Coord. Chem. Rev.* **178-180** 1433
[2] Ye R, Baba M, Ohishi Y, Mori K and Suzuki K 2006 *Mol. Cryst. Liq. Cryst.* **444** 203
[3] Bao Z, Lovinger A J and Dodabalapur A 1996 *Appl. Phys. Lett.* **69** 3066
[4] Zeis R, Siegrist T and Kloc C 2005 *Appl. Phys. Lett.* **86** 022103
[5] Law K Y 1993 *Chem. Rev.* **93** 449
[6] Mizuguchi J, Rihs G and Karfunkel H R 1995 *J. Phys. Chem.* **99** 16217
[7] Cornil J, Calbert J P and Brédas J L 2001 *J. Am. Chem. Soc.* **123** 1250
[8] Tada H, Touda H, Takada M and Matsushige K 2000 *Appl. Phys. Lett.* **76** 873
[9] Saito T, Sisk W, Kobayashi T, Suzuki S and Iwayanagi T 1993 *J. Phys. Chem.* **97** 8026
[10] Yonehara H, Ogawa K, Etori H and Pac C 2002 *Langmuir* **18** 7557
[11] Yonehara H, Etori H, Engel M K, Tsushima M, Ikeda N, Ohno T and Pac C 2001 *Chem. Mater.* **13** 1015
[12] Brinkmann M, Wittmann J C, Barthel M, Hanack M and Chaumont C 2002 *Chem. Mater.* **14** 904
[13] Hosada M, Wada T,Yamada A, Garito A F and Sasabe H 1991 *Japan. J. Appl. Phys.* **30** L1486
[14] Yamashita A, Maruno T and Hayashi T 1993 *J. Phys. Chem.* **97** 4567
[15] Tsuzuki T, Kuwabara Y, Noma N, Shirota Y and Willis M R 1996 *J. Appl. Phys.* **35** L447
[16] Norton J E and Bredas J L 2008 *J. Chem. Phys.* **128** 034701
[17] Tsamouras D and Palasantzas G 2002 *Appl. Phys. Lett.* **80** 528
[18] Sze S M 1981 *Physics of Semiconductor Devices* (New York: Wiley–Interscience)

AUTHOR INDEX